Pioneer of Intelligent Construction - Intelligent Compaction

Guanghui Xu · Dongsheng Wang ·
George Kuoneng Chang

Pioneer of Intelligent Construction - Intelligent Compaction

CHINA RAILWAY PUBLISHING HOUSE CO., LTD. Springer

Guanghui Xu
Department of Transportation Science and Engineering
Harbin Institute of Technology
Harbin, Heilongjiang, China

Dongsheng Wang
Department of Transportation Science and Engineering
Harbin Institute of Technology
Harbin, Heilongjiang, China

George Kuoneng Chang
The Transtec Group, A Terracon Company
Austin, TX, USA

ISBN 978-981-95-4183-6 ISBN 978-981-95-4184-3 (eBook)
https://doi.org/10.1007/978-981-95-4184-3

Jointly published with China Railway Publishing House Co., Ltd.
The print edition is not for sale in China (Mainland). Customers from China (Mainland) please order the print book from: China Railway Publishing House Co., Ltd.

The original submitted manuscript has been translated into English. The translation was done using artificial intelligence. A subsequent revision was performed by the author(s) to further refine the work and to ensure that the translation is appropriate concerning content and scientific correctness. It may, however, read stylistically different from a conventional translation.

This Springer imprint is published by the registered company Springer Nature Singapore Pte Ltd.
The registered company address is: 152 Beach Road, #21-01/04 Gateway East, Singapore 189721, Singapore

Preface

Compaction quality control is a crucial aspect in the construction of fill-type transportation infrastructure, including highways, railways, airports, and urban roads. Traditional compaction quality control is mainly "point-based." Continuous compaction control changes this mode, enabling continuous control throughout the rolling process, which is the "third revolution of road construction technology." Intelligent compaction, based on continuous control, integrates modern information technology and intelligent technology to achieve autonomous "perception, analysis, decision-making, control" and is hailed as the "fourth revolution of road construction technology." This book provides a concise and systematic explanation of the basic principles and latest development in intelligent compaction control technology.

This book consists of five chapters and two appendices, written according to the basic characteristics of intelligent construction—autonomous "perception, analysis, decision-making, execution." Chapter 1 introduces the characteristics and performance requirements of fill engineering, analyzes the characteristics of quality control indicators and traditional detection methods, provides a brief review of the history of compaction quality control methods, and provides an initial understanding of what intelligent compaction is. Chapter 2 is the theoretical basis of intelligent compaction. Starting from the characteristics of the rolling process, it introduces the mechanical principles of the interaction between the roller and the filler body, one of the "perception" foundations of intelligent compaction, analyzes the difficulty of solving and the method of identifying control indicators (modulus), and briefly explains the "perception" foundation. The second, continuous testing technology, analyzes the various factors affecting the test results. On this basis, it briefly introduces the basic characteristics of expert systems, neural networks, and reinforcement learning in the "analysis and decision-making" tool—the intelligent model. It then simply analyzes the core of the "execution" part, specifically the compaction control principle. Chapter 3 begins with the basic requirements of transportation infrastructure, focusing on the control elements during the rolling process, including the degree of compaction, compaction stability, and compaction uniformity. It provides their technical requirements and analyzes the coordination and cooperation issues of continuous control and point control. Chapter 4 primarily introduces the implementation steps of intelligent compaction, including the characteristics of the control system, the "perception" of compaction state, the "analysis" of compaction data, the "decision-making" of rolling pattern, the "execution" of rolling operation, and feedback control. To better understand intelligent compaction, an engineering example is provided at the end. Chapter 5 introduces some extended applications of intelligent compaction, including standardized continuous detection, construction robots, intelligent rollers, coordinated operation of machine groups, parameter stratification of fill bodies, and fusion of compaction data and observation data. Additionally, the role of intelligent compaction in virtual construction is also examined. Throughout the book, the authors have left some issues and directions worth exploring, delving into, and researching for young researchers in intelligent compaction and related technologies. If you pay close attention, you will find them.

Intelligent compaction, as a pioneer of intelligent construction, is a product of multidisciplinary integration—a concrete manifestation of the new generation of information technology in construction—and currently the most successful engineering application example. This book is the last volume of the "Frontier Technology Series of Intelligent Construction of Transportation Infrastructure" (First Phase) and is also the sister volume of "Dynamic Principles and Engineering Applications of Subgrade Continuous Compaction Control" (Science Press, 2016) and "High-speed Railway Subgrade Continuous and Intelligent Compaction Control Technology" (China Railway Publishing House, 2019). Although the length is not long, it systematically introduces the overall progress of intelligent compaction. During the writing of this book, Dr. George K. Chang, co-founder of the International Society for Intelligent Construction (ISIC) and the Transtec Group - A Terracon Company, provided materials and contributed to the writing work.

Readers are kindly requested to point out any issues in this book and the authors will correct them. Please send feedback to highx@163.com.

Harbin, China
Harbin, China
Austin, USA

Guanghui Xu
Dongsheng Wang
George K. Chang

Contents

Chapter 1
Fill and Control

Fill engineering is widely used across various infrastructures. Compared to bridge and tunnel engineering, it may seem simple, but it is actually quite complex. Its engineering quality is even harder to control. With the development of technology, both the method of fill compaction and the method of quality control have undergone profound changes. Intelligent compaction, as a new technology, integrates compaction and quality control of fill, fully embodying the four basic characteristics of intelligent construction: "perception, analysis, decision-making, execution," and is a pioneer of intelligent construction.

1.1 What Is Fill Engineering?

Fill engineering refers to various types of projects built by layering and compacting various construction materials. It is widely used in various infrastructures. Roadbeds and pavements in highways, railway roadbeds, airport pavements, urban roads, water conservancy dams, ports, and other similar structures all belong to fill engineering, as shown in Fig. 1.1.

Fill engineering has two notable features. First, the source of filler materials is complex. Various rock and soil materials, as well as man-made mixed materials (such as concrete, asphalt mixtures, secondary ash, and crushed stone), construction waste materials, and tailings materials from mines, can all be used as fill materials. Second, it isn't easy to control the quality of the project. Due to the complexity of the fill material, there is generally an uneven distribution of particle shape, gradation, moisture content, etc., which makes the representativeness of the sampling test results of the filler body poor and makes it difficult to fully reflect the project quality.

The line part (subgrade and pavement) of transportation infrastructure falls under the field of filling engineering, which is a "linear" structure. Its engineering quality is the most difficult to grasp among the three structures of "roads, bridges, and tunnels",

G. Xu et al., *Pioneer of Intelligent Construction - Intelligent Compaction*,
https://doi.org/10.1007/978-981-95-4184-3_1

Fig. 1.1 Fields involved in fill engineering

and it has the characteristics of concealed engineering. As a filler, under certain filling conditions, the focus is on how to carry out filling compaction and quality control, which is the core issue in the construction of the line (filler engineering) and has always been the focus of the engineering construction field and the engineering machinery field. It is also the focus of this book. For convenience, unless specifically stated, the fill mentioned in this book refers to the subgrade and pavement in roads and railway engineering. Of course, the content of the book is also applicable to other engineering fields related to fillers, such as airport engineering and urban roads.

1.2 Performance Requirements of Fill

The "line" is a fill-type structure (also called a fill or filler), which accounts for the largest proportion of transportation infrastructure. Its main task is to support the repeated actions of traffic loads (cars, trains, and airplanes) and provide a comfortable and safe operating environment for vehicles. Therefore, the performance of the fill mainly encompasses four aspects: strength, stiffness, stability, and uniformity, all of which must meet the specified requirements, as shown in Fig. 1.2.

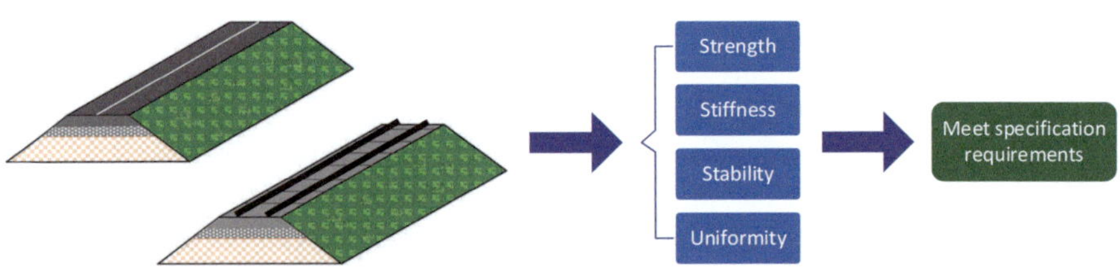

Fig. 1.2 Performance requirements of fill

(1) **Strength**

Strength refers to the ability of the fill to resist damage (rupture or excessive plastic deformation) under the repeated action of vehicle load and natural factors (water and temperature). This is an indicator of the damage to the fill, that is, the ability of the fill to resist external damage, generally represented by allowable stress [σ].

During construction, the fill's strength is formed under the action of roller compaction, which is a compaction action where the plastic deformation of the fill changes from large to small until it reaches zero. From another perspective, the compaction process involves the destruction of the old fill structure and the continuous formation of new structures, resulting in a gradual increase in strength until it meets the specified requirements.

What kind of strength requirements should a built-fill meet? This strength requirement is determined through tests, typically using a direct shear test or a triaxial shear test. During construction, it is impossible to conduct such field tests; however, this requirement is indirectly met through quality control.

(2) **Stiffness**

Stiffness refers to the ability of the fill to resist elastic deformation under the repeated action of vehicle load and natural factors (water and temperature). Under the same load, the stiffer fill has a larger corresponding strength, but exhibits smaller elastic deformation.

The stiffness reflects the elastic deformation properties of the fill and is essentially a measure of the resistance to elastic deformation, as reflected through the stress–strain relationship. It is generally characterized by the modulus of elasticity—the force resisting elastic deformation per unit area, measured in Pa (N/m^2) or MPa (MN/m^2, N/mm^2). Under the same stress conditions, the stiffer fill has a smaller elastic deformation. Figure 1.3 shows the stiffness and corresponding strength of different fills.

Stiffness is a crucial mechanical property of the fill, determined by the method of connection between the fill particles. When the connection between the particles is strengthened, the ability of the filler to resist deformation also increases, and the

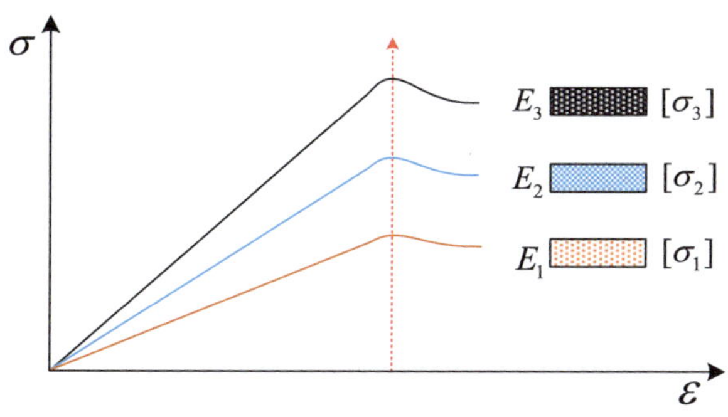

Fig. 1.3 Stiffness and strength of different fills

stiffness is greater, and vice versa. Stiffness is a measure of the fill's deformation, and controlling the stiffness also controls the elastic deformation to a certain extent.

There are clear requirements for the stiffness of the fill. Generally, the technical standards specify the stiffness requirements for commonly used fill materials, such as sandy soil fill, which requires the modulus $[E] \geq 60$ MPa, which is also the compaction standard.

Stiffness can be determined by a plate load test. Due to different loading and processing methods, multiple indicators, such as K_{30}, E_{v2}, and E_{vd}, can all be used to characterize stiffness. Although theoretically, stiffness is characterized by the modulus of elasticity, due to the complexity of the fill, multiple expressions are allowed.

It should be noted that whether it is stiffness or strength, it is important to distinguish between the properties of the fill itself and the properties of the structure it forms. For example, in large-diameter stones, their stiffness and strength are greater than that of the fill structure formed. The two should not be confused.

(3) **Stability**

Stability generally refers to the filler body's ability to maintain its original equilibrium state under the repeated action of vehicle loads and natural factors (water and temperature), that is, the ability of the filler body to resist deviation from that state.

The state of the filler body (from the system's point of view) is typically characterized by its physical mechanics, such as its modulus and density. It can also be represented by stress and strain. Therefore, stability usually refers to the property that the physical and mechanical properties of the filler body do not change significantly with time.

In a broad sense, the stability of the filler body encompasses aspects such as water stability and temperature stability. In transportation infrastructure, the fill structure is subjected to repeated loading from vehicles, as well as the effects of the natural environment (water and temperature). Maintaining sufficient stability is crucial and a prerequisite for improving engineering quality.

If the performance of the filler body does not change significantly throughout its life cycle, it means that its structure is stable and the engineering quality is high, as shown in Fig. 1.4.

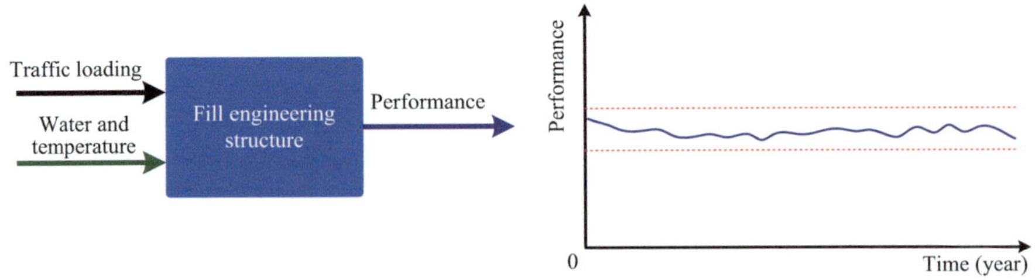

Fig. 1.4 Schematic diagram of the relationship between filler body performance and time

If the stability of the filler body decreases during its service life, it indicates a decrease in various performances (such as stiffness). Apart from early damage causes, the most likely reason is fatigue—the structure will be damaged under conditions below its strength, which is the inevitable result of the long-term, repeated action of vehicle loads.

The evaluation of the stability of the filler body requires long-term monitoring. Accelerated loading tests are a quick evaluation method, but they are different from the real state. Additionally, this type of test is time-consuming and labor-intensive, and the test equipment is very expensive, making it challenging to integrate into the construction process. Therefore, during the compaction stage of the filler, the compaction quality should be well controlled, which, in turn, indirectly controls stability.

The filler body has sufficient strength, stiffness, and stability, which are the basic requirements for vehicle driving. In addition to these requirements, for "linear" projects such as highways and railways, the filler body must also provide uniform support for vehicles, which is the issue of uniformity.

(4) **Uniformity**

Uniformity refers to the consistency of the distribution of the physical and mechanical properties of the filler body in the horizontal direction. This is a special requirement for line projects that stretch for tens or even hundreds of kilometers.

If the performance of the filler body is unevenly distributed on the plane, it means that the strength and stiffness have varied. Under repeated vehicle loads, due to the relatively low stiffness and strength of the weak areas, larger deformations will occur under the same load. In severe cases, shear failure (settlement, permanent deformation) may occur, posing a safety hazard to the upper structure and driving, as shown in Fig. 1.5.

The non-uniformity (variability) of the filler body performance is readily apparent in structures built with coarse aggregate and mixed materials. There are two reasons: one is that the fill material is segregated during spreading, causing the separation of coarse and fine materials, which is the fundamental reason; the second is that the same number of compaction passes and compaction processes are used to compact uneven fill materials, the result must be uneven, so the number of compaction passes should be optimized, and different numbers of compaction passes should be used in different areas.

As for the quantitative requirements for the uniformity of the filler body performance, it remains relatively challenging to determine at present, and further research is still needed from both theoretical and practical perspectives. The size of uniformity

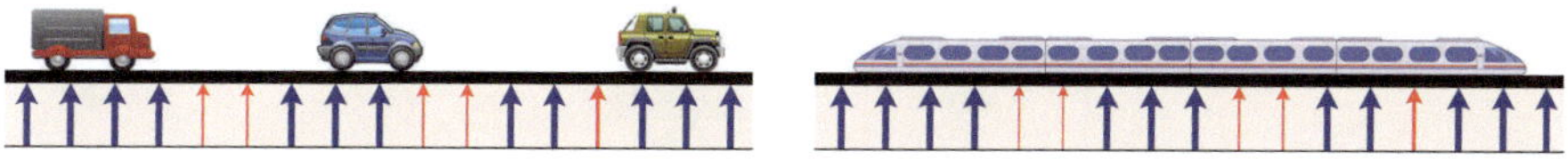

Fig. 1.5 Uneven distribution of filler body performance affects driving safety

is related to the requirements of the upper structure and driving, and is now mostly determined by empirical methods.

1.3 Evolution of Construction Methods

Since the advent of humans, roads have existed, evolving from ancient dirt and gravel roads to structural-laid stone roads, and ultimately to modern asphalt and concrete roads, which have endured for hundreds of years. In contrast, railways only emerged in the mid-nineteenth century. For fill construction, whether it is a road or a railway (except for the track), the construction process is the same: lay a certain thickness of fill material and then compact it. With the advancement of science and technology, the construction method (centered on compaction) has evolved from ancient manual tamping to modern intelligent compaction, as illustrated in Fig. 1.6.

Traditionally, the technology of building highways and railways is referred to as "road construction technology." Generally, the evolution from manual tamping to mechanical static compaction is called the "first revolution" of road construction technology; mechanical vibration compaction is called the "second revolution"; continuous compaction control road construction technology is called the "third revolution"; the developing intelligent compaction will be a new revolution in road construction technology. Currently, there are three main compaction methods corresponding to different construction machinery, as shown in Fig. 1.7.

The main machinery for static compaction includes light drum rollers and rubber-tired pneumatic rollers, also known as rolling or static pressure rollers, which uses static force to compact the fill. Static compaction is mainly suitable for fine-grained materials, but can only compact the filler to a certain extent. To further improve compaction, you must modify the compaction method. Corresponding to static

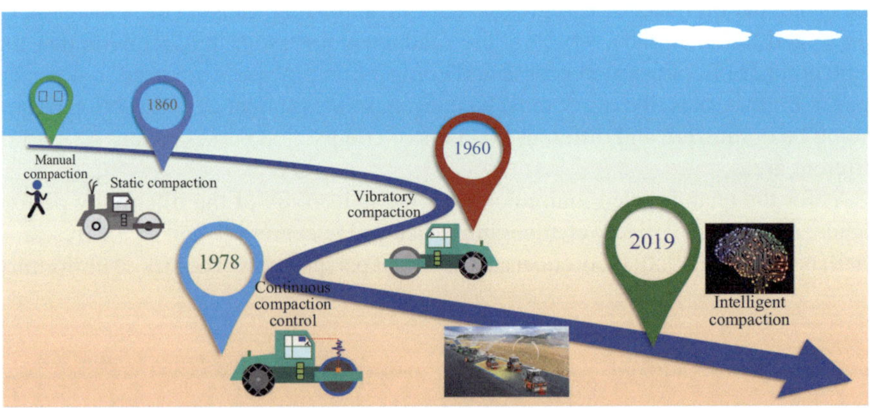

Fig. 1.6 Evolution of highway and railway construction methods

Fig. 1.7 Current compaction methods and compaction machinery

compaction is dynamic compaction, which is divided into impact compaction and vibration compaction.

Impact compaction is also known as tamping. Originally, it was done by using a piece of iron that fell freely from a certain height, relying on the generated impact force to compact the fill. The range covered by impact compaction is wide, encompassing various mechanical impact rollers and other tools.

Vibration compaction is another type of dynamic compaction method, characterized by a large compaction depth and the most extensive application. This technology originated in the 1930s, matured in the 1960s, and is considered the "second revolution" in road construction technology. Vibration compaction applies a periodic vibration load to the surface of the fill, causing the fill particles to transition from a stationary state to a repetitive motion state, ultimately to a stable position. This process enables the fill particles to interlock and achieve a relatively compact, stable state, thereby strengthening the particle interlock. Another type of vibration compaction is horizontal vibration, also known as oscillating compaction, which is suitable for thin-layer rolling.

Vibration compaction is widely used due to its high compaction efficiency and broad applicability. It is currently the main compaction method for filler construction. Continuous compaction and intelligent compaction are mainly achieved using vibration rollers.

1.4 Quality Control Indicators

The fill structure must meet the requirements of strength, stiffness, stability, and uniformity in performance. How does the performance of the filler meet these requirements? This is the main task during construction. Through quality control, certain indicators are monitored to ensure they meet the specified requirements, indicating that the performance of the fill has also met these requirements. Therefore, choosing what kind of control indicators is crucial. The selected indicators should characterize the performance of the fill, be easily detectable on-site, and account for the complexity of the fill in the control indicators.

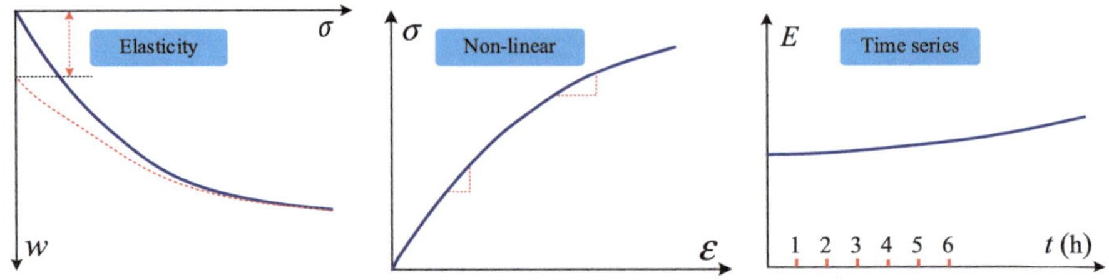

Fig. 1.8 Manifestation of the complexity of the fill

(1) Complexity of the fill

The fill materials for building roads and railways come from a variety of sources. The subgrade part is mainly composed of rock and soil materials. Due to the characteristics of rock and soil materials, the indicators characterizing the performance of the fill are proposed relative to a certain state and are conditional indicators. These indicators undergo complex changes in response to external actions (such as load, temperature, and humidity) and are not constant. This complexity is mainly manifested in aspects such as elasticity, nonlinearity, and time effect, as shown in Fig. 1.8.

First, the complexity of the filler is evident in its elasticity. In principle, the fill that is rolled and formed should be in an elastic state, but it is not achievable in practice. They are all elastoplastic bodies. This is evident from the results of the plate load test. There is a certain amount of plastic deformation, which gives rise to the concept of deformation modulus and rebound modulus.

Secondly, the complexity of the fill is reflected in the nonlinearity of the stress–strain (constitutive) relationship. There is an essential difference between linearity and non-linearity. There is only one kind of linear relationship—a straight-line relationship, but there are countless kinds of non-linear relationships. The actual fill rarely completely conforms to the linear elasticity theory. Even if it is elastic, it is also non-linear. That is, the stress–strain relationship is a curve, not a straight line.

The elastic modulus of an ideal linear elastic body is of constant value and does not change with external changes. However, a large amount of experimental data indicates that the indicators obtained by different test methods and loading methods vary. For example, using the same plate load test method, due to different loading requirements, different moduli will be obtained, including rebound modulus (E), first deformation modulus (E_{v1}), second deformation modulus (E_{v2}), and so on.

The non-linear characteristics of the fill make the performance indicators conditional indicators; they change with different test methods and loading methods, introducing some uncertainties into compaction quality control. For example, after rolling is completed, different loading levels are used for the plate load test, resulting in different outcomes. Using different test methods yield results are even more varied, and sometimes, they may even be contradictory. Typical examples are E_{vd}, E_{v2}, K_{30}. Although they can all reflect stiffness, their values differ, and the results are not entirely consistent (commonly referred to as "indicator" conflicts).

The current level of science and technology is still largely focused on linear thinking and problem-solving, but the world is essentially nonlinear. For filler bodies, the current design and calculation are based on linear elasticity. For example, the theory of drum-rail dynamics assumes the filler body is linear, and models it as a simpler "Winkler" foundation or spring model. Road mechanics are also calculated as a layered elastic body. However, the actual situation is not the case; otherwise, there would not be so many problems.

Finally, the complexity of the filler material is also reflected in timeliness—various performance parameters will change over time. A typical example is that the modulus of the filler body changes with the seasons. In addition, after the filler body is rolled and formed, the results of the immediate plate load test differs from those after a few hours (or even longer). The latter is generally greater than the former. The same problem exists in continuous testing with a vibrating roller. After the rolling is completed and the test is carried out a few hours later, the result will change, indicating that its compaction state has changed. That is, its physical and mechanical properties have changed. This is a phenomenon unique to geotechnical materials and cannot be avoided.

The performance parameters of the filler body are nonlinear, time-dependent, and variable due to the complexity of the fill material itself (internal factors) and the variability of external actions (external factors), which complicates the selection and application of quality control indicators. It is impossible to avoid it completely; controlling and adapting are problems worth studying, and perhaps new structural forms need to be developed to better adapt to this complexity.

(2) **How to choose quality control indicators**

During the rolling process, the type of indicators to choose for controlling compaction quality is critical. The principle of selecting indicators is straightforward: the indicators should accurately reflect the performance characteristics of the filler body. Essentially, whether it is a highway, railway, city road, or airport, it is a mechanical problem that supports the smooth operation of vehicles and the take-off and landing of aircraft. Therefore, the purpose of compaction is to improve the ability of the filler body to withstand traffic loads (including water and temperature). Therefore, when choosing control indicators, the first consideration should be mechanical indicators.

The filler body is a continuum, and the current models are based on elasticity mechanics. Continuum mechanics has three physical parameters—modulus, Poisson's ratio, and density—that represent its mechanical properties (modulus and Poisson's ratio) and its physical properties (density). If it is a dynamic model, the modulus, Poisson's ratio, and density are required for calculation. If it is a static model (with acceleration set to 0), the density does not affect the calculation. Therefore, the first choice for compaction quality control indicators is the modulus, followed by density.

Modulus and density are two independent indicators, that respectively reflect the mechanical properties and physical properties of the filler body. It is the correct choice to use these two indicators for compaction quality control. The modulus can be used

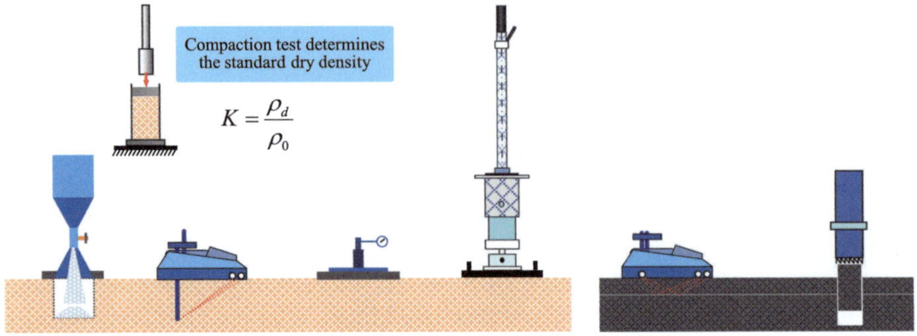

Fig. 1.9 Schematic diagram of detection methods

directly as a control indicator, and the density appears in the form of a compaction coefficient (also known as compaction degree, expressed as a percentage)—the ratio of the field dry density to the standard dry density (obtained through an indoor compaction test).

(3) **Field detection method**

After selecting modulus and density as the compaction quality control indicators, the next step is to perform the measurement. The traditional measurement method randomly selects detection points on the site, which is a "point" control. Figure 1.9 shows the schematic diagram of several commonly used detection methods.

For the modulus, the plate load test method is usually used for detection. Using different loading methods and calculation methods, various specific indicators are derived, including the rebound modulus (E), deformation modulus (E_{v2}), and Winkler foundation reaction modulus (K_{30}), among others. Among them, the rebound modulus test takes the longest time and requires repeated loading. To eliminate the impact of plastic deformation, the test time is at least 1 h. The measurement time of the Winkler foundation reaction modulus is the shortest, approximately 20 min. Another quick detection method is the LWD (Light Weight Deflectometer). The indicator obtained is E_{vd}, which is a dynamic modulus; however, its accuracy is not as high as that of the plate load test, which is commonly used for process control during rolling.

Regarding density, rock and soil fillers are detected using the sand cone method and the densitometer method. It is necessary to measure the onsite dry density and compare it with the standard dry density to determine the compaction factor. The detection time of the sand-cone method is also relatively long, requiring tens of minutes. The densitometer method is an auxiliary means that has evolved from the nuclear densitometer to the non-nuclear densitometer. The detection time is relatively short, but it requires calibration in advance, and the accuracy is limited. It is commonly used for controlling the compaction process. The core extraction method is commonly used on asphalt pavements and semi-rigid base, and is a destructive test. It requires the sample to be taken back to the laboratory for further testing, which is quite troublesome. The density index is mainly applicable to cohesive fine aggregates (cohesive soil that is sensitive to water). For coarse aggregates, it is very difficult to use the density index for compaction quality control because its accurate measurement is very difficult.

1.5 Advantages and Disadvantages of Traditional Control Methods

Traditional compaction quality control is completed through on-site "sampling point" detection. This type of "point-based" control method is familiar to everyone, and its advantages and disadvantages are obvious.

(1) **Advantages of point-based control methods**

Plate load tests and sand filling methods are traditional methods for detecting and controlling modulus and dry density. They are standard methods stipulated in the technical standards of various countries, suitable for quality acceptance, and are also accepted by all walks of life for their detection accuracy. All other detection methods must be verified using these two methods.

Since the plate load test and sand cone method (including the core extraction method) are standard methods stipulated in the technical standards of various countries, it is unrealistic to cancel them completely at present (construction units have this idea). They will continue to exist until new detection methods mature.

(2) **Disadvantages of point-based control methods**

The most obvious disadvantage of traditional detection methods is the lengthy test time, making them time-consuming and laborious. The plate load test also requires the use of large construction machinery as a reaction load device. In addition, these detections are generally carried out after the compaction is completed, which is a post-detection step, and problems need to be reworked.

Figures 1.10 illustrate the on-site operation process of these methods. It can be observed that the labor intensity is high, the time consumption is prolonged, the number of tests that can be completed in a day is limited, and this approach cannot meet the quality control needs of mechanized construction.

Fig. 1.10 Time-consuming and laborious point-based detection methods

Time-consuming, labor-intensive, and post-detection are the main disadvantages of traditional detection methods, problems inherent to the test method itself. In addition, when using these "point-based" detection results for compaction quality control, there are other important issues (such as the impact of filler particle size) that we hope will also attract attention.

1) **Whether the point-based control results are representative**

"Point-based" control is based on the detection results of "sampling points." Although modulus and density can be obtained, it is worth discussing whether the data from a few sampling points can accurately represent the compaction quality of the entire compaction surface.

The sampling method originates from statistics, where sample data are used to evaluate a population, but it must meet certain conditions. For fill compaction, the "sampling points" on the compaction surface are samples, and the entire compaction surface is the population. Whether the data from the sampling points accurately represent the compaction quality of the compaction surface depends on the distribution of the compaction state across the compaction surface. Figures 1.11 present the conditions for the sampling method: all data on the compaction surface (such as modulus) must conform to a normal distribution, and sampling is meaningful at this time.

A large amount of practical experience indicates that when the filler is relatively uniform, all compaction quality data on the compaction surface indeed conform to a normal distribution, and point-based control is feasible.

Suppose there is unevenness (variability) in the filler. In that case, the overall compaction quality data on the compaction surface does not conform to a normal distribution (as shown in the lower figure of Fig. 1.11), and the data from the sampling points cannot represent the entire compaction surface. In this case, the point-based control method is invalid.

Suppose every point on the compaction surface can be detected in real-time. In this case, it is undoubtedly the best for quality control, allowing problems to be addressed in a targeted manner in a timely manner. "Area control" adopts a different

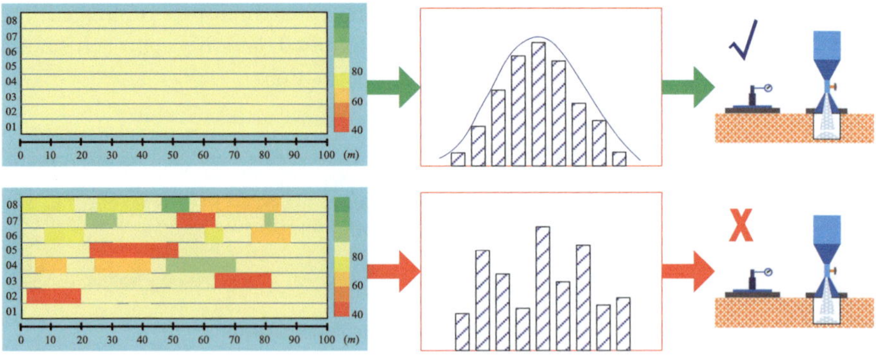

Fig. 1.11 Applicable conditions of the sampling method

technical approach compared to "spot test control." It is not only comprehensive but also fast in detection. See the introduction later.

2) **The problem of standard dry density**

The original standard dry density (compaction standard) was proposed for cohesive soil. This material is sensitive to water. The standard compaction test (Fig. 1.9) can be used to obtain the relationship between water content and dry density. At a certain water content, the dry density is the largest, and the mechanical properties are also the most stable. This water content is referred to as the optimum water content. During compaction, it is necessary to strictly control water content to keep the filler within the allowable range of the optimum water content, thereby achieving the best compaction.

The impact test was later applied to other materials, such as soil and stone mixtures, but numerous problems arose. The biggest problem is that variation in fillers leads to different optimal moisture contents and maximum dry densities across different samples, and the compaction standards used differ as well. The site control does not know which test result to choose, which can easily confuse quality control.

3) **The relationship between modulus and density**

Modulus and density are two independent indicators reflecting the mechanical and physical properties of the filler body, respectively. For a purely elastic body, there is no relationship between the two in theory. Otherwise, they would not be independent. However, in practice, it has been found that, under certain circumstances, a positive relationship exists between modulus and density. For the same material, the more compact the pressure, the larger the modulus of the formed structure (Fig. 1.12). Therefore, when controlling compaction quality, some people advocate canceling the compaction coefficient and retaining only the modulus, which is one of the reasons.

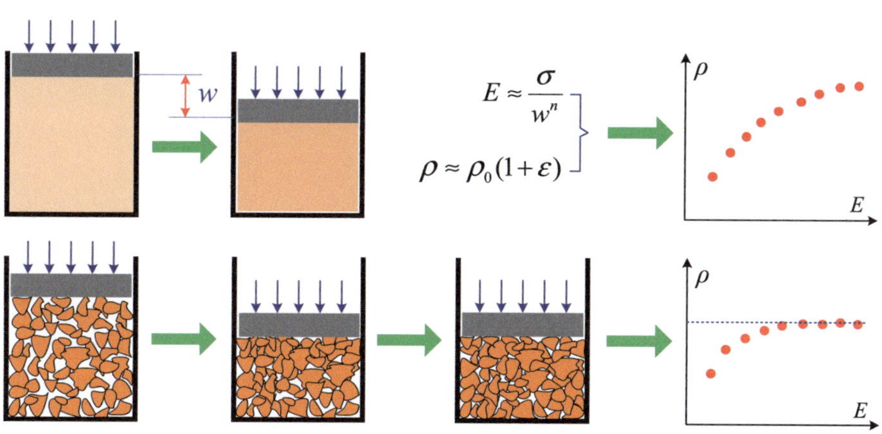

Fig. 1.12 The complex relationship between modulus and density

For cohesive soil, if the moisture content is controlled within the optimal moisture content range, there is indeed a strong correlation between density and modulus; if the moisture content distribution is wide, the two may not necessarily be correlated. The modulus is the largest when the moisture content is small, but the dry density is not. If controlled only by modulus, although the mechanical properties of the filler body can be guaranteed, instability will occur when it encounters water, and its mechanical properties will decrease. Therefore, for the compaction control of cohesive soils (or water-sensitive materials), the density index remains very useful, as it can control water stability. This is why it is necessary to control the optimal moisture content.

For coarse-grained materials, especially block stone fillers, the situation is more complicated. When the density stops increasing, the modulus may continue to increase (Fig. 1.12), and there is no clear correlation between the two. The key element determining the mechanical properties of the filler body is the degree of association between particles (the size of the internal friction angle), not the degree of compaction. Therefore, for the rolling of coarse-grained materials, the control is mainly based on modulus.

——The compaction coefficient (compaction degree) is a long-controversial indicator. The focus is on whether this indicator can reflect the compaction quality. The practice has shown that for water-sensitive materials (such as cohesive soils), controlling compaction quality to the maximum dry density (as per the compaction standard) ensures the most stable mechanical properties of the filler body. For other fillers, the relative size of the density (compaction coefficient) cannot reflect the mechanical properties of the filler body. The primary reason for choosing density is that it is much simpler to measure density on-site than to measure modulus. Many people believe that the denser the filler body, the stronger its ability to resist deformation is, and they try to characterize the mechanical properties by the relative size of the filler body's density (compaction coefficient), but this is unreliable. In the technical standards of Chinese railways, only the control of the compaction coefficient is required for cohesive fine-grained materials. Various moduli are used to control the compaction quality of coarse-grained materials.

1.6 From Continuous Compaction to Intelligent Compaction

For fill construction, under certain fill conditions, compaction quality control is the key link in determining the performance of the filler body, which is directly related to the ride quality of vehicles and ensures comfort and safety issues. The advantages and disadvantages of traditional compaction quality control methods have been analyzed above. Although there is no problem with the control indicators themselves, there are many problems with the detection methods. The detection process is lengthy and laborious, and the results are not sufficiently representative, which is a significant disadvantage.

In response to the shortcomings of traditional control methods, a technology for controlling compaction quality using roller's dynamic response information emerged

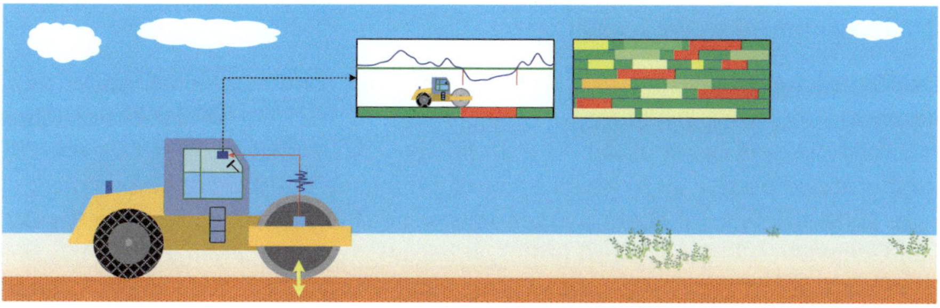

Fig. 1.13 The working process of continuous compaction control

in Northern Europe in the 1970s— Continuous Compaction Control (CCC)—which replaced the traditional point control.

Continuous compaction control (referred to as continuous compaction) refers to the establishment of a detection evaluation and feedback control system during the rolling process of the fill, based on the principle of interaction between the roller and the filler body, by continuously measuring the response signal of the roller's vibrating drum, to achieve real-time dynamic monitoring and control of the compaction quality of the entire rolling surface. Figure 1.13 shows the working process of continuous compaction.

Continuous compaction is a general term for a type of "surface control" method, and there are many specific methods. It has now progressed from continuous compaction to intelligent compaction. The following provides a brief introduction to the development situation.

(1) **Development in Europe**

Sweden developed a measuring device called a "compaction meter" in 1976, which continuously measures the ratio of the base frequency and the first harmonic amplitude of the vibration response signal of the roller's vibrating drum (called compaction meter value or CMV) during the rolling process. It was first applied in construction in 1978. After entering the 1980s, other Nordic countries also conducted research and development based on the same CMV principle. Since the 1990s, some Nordic countries have gradually begun to formulate technical standards for continuous compaction, emphasizing that the continuous compaction index should be correlated with traditional point-based compaction results (compaction coefficient and deformation modulus). In addition to Sweden's requirement of a correlation coefficient of at least 0.6, other countries require a correlation coefficient of 0.7 or higher.

After the year 2000, Germany and Switzerland began using mechanical indicators (modulus and stiffness coefficient) as control indicators for continuous compaction. They developed vibrating rollers with automatic frequency and amplitude adjustment. There has been great progress in measurement equipment, and a complete compaction control and management system has emerged. Some countries have successively updated the continuous compaction technical standards.

(2) **Development in the United States**

The United States began researching and applying continuous compaction technology in 2000, using satellite positioning to achieve precise control over the rolling track and position of the roller. This technology is referred to as "CCC + GPS" intelligent compaction (IC), which is currently in its initial stage of development.

In 2004, the Federal Highway Administration (FHWA) announced the "Intelligent Compaction Plan (IC Road Map)". It began promoting intelligent compaction technology, primarily used in asphalt pavement rolling control, focusing on controlling the number of rolling passes and the temperature. In 2007, the FHWA, the Transportation Pooled Foundation (TPF), and the National Cooperative Highway Research Program (NCHRP) initiated scientific research and the application of this technology, as outlined in the TPF-5 (128) research plan and the national promotion plan, Everyday Counts (EDC). In 2012, FHWA issued intelligent compaction standards for soils, subbase and asphalt mixtures. In 2014, the American Association of State Highway and Transportation Officials (AASHTO) issued national standards for intelligent compaction. In 2017, FHWA published an official document on the grading of intelligent compaction technology (FHWA-HIF-17-046) (this document was initially drafted by the ISIC Executive Committee), clarifying that technologies above level 3 should adopt mechanical control indicators, promoting the development of intelligent compaction technology towards a higher level. In the same year, the Artificial Neural Network (ANN) technology was used in the prediction of compaction control indicators, starting the integration with artificial intelligence, and gradually began to shift the control indicators from empirical indicators (such as CMV) to mechanical indicators (modulus), and studied the algorithm for stratifying the performance parameters of the filler body (such as NCHRP 24-45 research). From 2015 to 2020, the TPF Intelligent Compaction Implementation Project (TPF-5 (334)) developed the national intelligent compaction standard software Veta, which became the standard of the Federal Highway Administration, accelerating the standardization and implementation of intelligent compaction in the United States.

To date, 25 states in the United States have issued intelligent compaction technical standards, making this technology widely adopted. The Minnesota Department of Transportation began to require all asphalt road construction to use intelligent compaction in 2018. The Missouri Department of Transportation began to implement intelligent compaction for asphalt pavement construction in 2017. Therefore, the United States has become one of the largest markets for intelligent compaction technology globally.

(3) **Development in China**

In 1988, the Chinese hydropower sector began using compaction meters for earth dam rolling control. Subsequently, the highway department introduced compaction meter technology, and some electronics manufacturers (including roller manufacturers) began producing measurement equipment based on the CMV principle. However, due to the inherent limitation of CMV, the correlation with the traditional point detection results is poor. In many cases (especially coarse-grained fill), CMV does not accurately reflect compaction quality, leading people to generally believe that this technology is not feasible and that the application has entered a trough.

In 1993, the Harbin Institute of Technology initiated a scientific research project on highway gravel soil subgrade construction control technology in Heilongjiang Province, marking the beginning of its independent research efforts. This was the first continuous compaction scientific research project funded by the government in the highway field. In 1998, the project team proposed a dynamic method with the subgrade structure's anti-deformation ability as the control indicator. It began experimental application in the highway field of the three northeastern provinces. In 2004, the Ministry of Communications initiated a research project on asphalt pavement construction quality control, applying continuous compaction technology. In 2009, Southwest Jiao-Tong University (primarily the original HIT research team) undertook the continuous compaction research project of the Ministry of Railways. It began systematic research on the application of this technology in high-speed railway construction. In 2011, the Ministry of Railways issued China's first industry construction standard for continuous compaction technology, requiring a correlation coefficient of at least 0.7 between continuous detection indicators and point detection indicators. In the same year, the National Natural Science Foundation funded the in-depth research of this technology. In 2015, China Railway Corporation issued a new standard for continuous compaction technology. In the same year, China proposed calculation methods for the resistance coefficient and estimated modulus, as well as the concept of parameter stratification based on the roller's compaction thickness. In 2017, the Ministry of Transport of China issued the industry product standard for continuous compaction technology, clarifying that mechanical indicators must be used for continuous compaction control. From 2022 to 2023, Heilongjiang Province and the China Highway Society respectively issued local standards and group standards for highway intelligent compaction technology, once again clarifying that the control indicators should be mechanical quantities (modulus is the best), and for the first time added "perception, learning, analysis, decision-making, control" and other intelligent technology content.

(4) **Development in other countries**

Australia began implementing engineering applications in 2018 and began developing relevant technical standards. Although its development began later than that of Europe, the United States, and China, it has progressed rapidly. As for Japan, the industry discovered the limitations of the CMV in the 1990s and made adjustments.

The above is a brief history of the development of continuous compaction technology. Currently, the countries and international organizations that have issued technical standards include Sweden (Highway Administration), Germany (Department of Transportation), Austria (Federal Highway Administration), Finland, International Society for Soil Mechanics and Geotechnical Engineering (ISSGE), the United States (FHWA, NCHRP, AASHTO, various state transportation departments), China (former Ministry of Railways, China Railway Corporation, Ministry of Transportation, Hebei Province and Heilongjiang Province, China Highway Society). These technical standards have laid the groundwork for the promotion and widespread

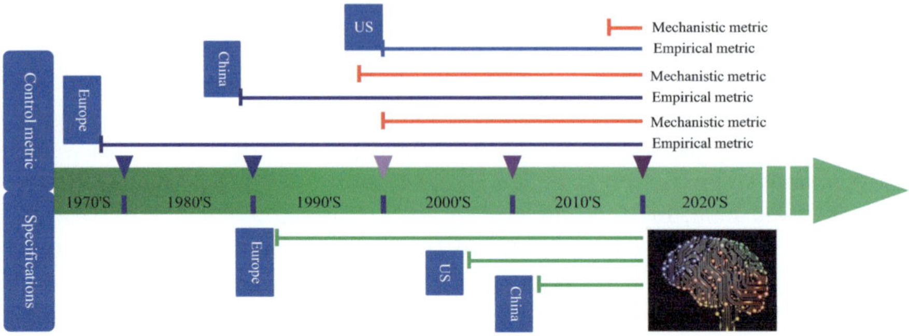

Fig. 1.14 Brief history of continuous compaction development

adoption of this technology worldwide. Figure 1.14 presents the main developments in continuous compaction technology and standards in a graphical format.

In 2016, China, the United States, and the European Union jointly initiated the establishment of the International Society for Intelligent Construction (ISIC), actively advocating and promoting new concepts and technologies for intelligent infrastructure construction globally and continuously deepening the connotation of intelligent compaction. At the second International Intelligent Construction Technology Conference in 2019 (Beijing), the ISIC Executive Committee proposed the basic requirements for achieving advanced intelligent compaction, enabling the roller to possess the abilities of "perception, learning, analysis, decision-making, and control." This has promoted the transition from continuous compaction to intelligent compaction, ushering in a new era of compaction quality control. Next, let's understand what intelligent compaction is.

1.7 Introduction to Intelligent Compaction

Intelligent compaction is a shorthand term for intelligent compaction control, a new technology developed based on continuous compaction. It inherits the advantages of continuous testing and has some characteristics of modern information technology (including artificial intelligence). Since the tool for intelligent compaction is a roller (currently mainly vibratory rollers), the implementation of intelligent compaction is centered around the roller. Many rollers now have the characteristics of informatization and networking, which lays a good hardware foundation for achieving intelligence.

(1) **Informatization, networking, and intelligence of rollers**

With the advancement of information technology, the Internet of Things has emerged as a new trend in the evolution of network technology. IoT technology is increasingly being applied in engineering construction, and the development of "information

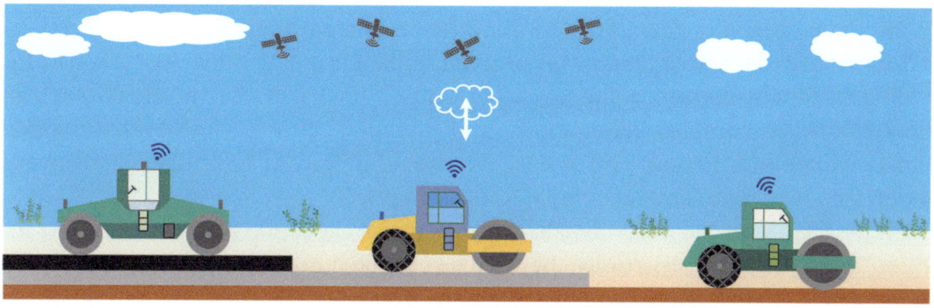

Fig. 1.15 Informatization and networking of rollers

construction sites" is rapid. As the main body of construction machinery, the roller is now generally equipped with microcomputers (computers), achieving "computerization" or informatization. On this basis, networking can be easily achieved, and with satellite positioning technology, remote monitoring of the operating conditions and travel positions of the roller have been realized, as shown in Fig. 1.15.

After the roller achieves informatization and networking, we can manage the roller remotely through the information platform. However, these alone are not enough. The roller should be made smarter; it should have the ability to self-perceive operating status and fault diagnosis (performance intelligence), and it should also self-diagnose and provide feedback control of compaction operations (functional intelligence). The former belongs to the category of intelligent manufacturing, and the latter is the task of intelligent compaction. The prerequisite for achieving these aspirations is first to realize the intelligence of the roller.

The operation of making an object intelligent is simple: to equip it with computer hardware, software, algorithms, and corresponding knowledge to form a control system. Among them, hardware and software (referring to the control software for the equipment operation part) exhibit a certain universality, while algorithms need to be designed for specific problems and implemented into the control system. Figures 1.16 show the ways in which the roller's intelligence is realized.

Equipping a roller with a computer is equivalent to implanting a "brain" into it. But if it does not have "thoughts," it can only be said to have achieved informatization and networking. The roller should also have "thoughts," replacing part of human mental labor, which is intelligence.

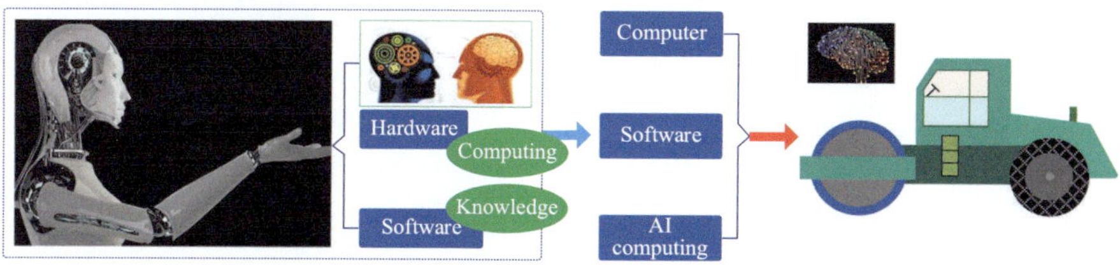

Fig. 1.16 Ways to realize the intelligence of rollers

Giving a computer the ability to think like a human brain has always been the dream of artificial intelligence (AI). Although the current AI does not possess the intelligence of a human brain, it outperforms humans in certain areas, such as repetitive tasks, complex calculations, and literature retrieval, and even surpasses human capabilities in some instances. We should utilize AI's strengths effectively, understand its nature, and leverage it to our advantage.

Additionally, the roller should also have a precise execution mechanism. After the roller "brain" issues instructions, they can be executed immediately, which is a form of feedback control. Rollers with automatic frequency and amplitude adjustment functions undoubtedly benefit the realization of advanced intelligent compaction and intelligent rollers.

(2) The meaning of intelligent compaction

In engineering construction, we care more about the intelligence of roller functions. Based on the interaction between the roller and the filler body, autonomously perceive compaction quality information and make intelligent diagnosis and feedback control of compaction operations. This is the task of intelligent compaction.

Although there is no clear definition of intelligent compaction at present, the basic meaning is clear. Figure 1.17 illustrates the working process of intelligent compaction, which helps explain what it is.

From the working process of intelligent compaction, it can be known that intelligent compaction refers to the process of obtaining compaction quality control information based on the continuously perceived vibration drum response signal of the roller during the compaction process and through autonomous learning of control information, filling information and compaction process information, it realizes autonomous analysis, decision-making and feedback control of compaction quality, thereby improving compaction quality.

(3) Intelligent compaction control system

Intelligent compaction is realized through a control system embedded in the roller. This control system can be divided into three subsystems, as shown in Fig. 1.18. The measurement system is responsible for perceiving the interaction information

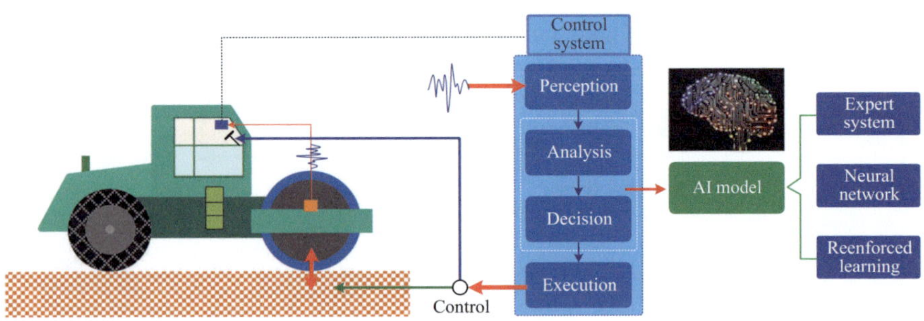

Fig. 1.17 The working process of intelligent compaction

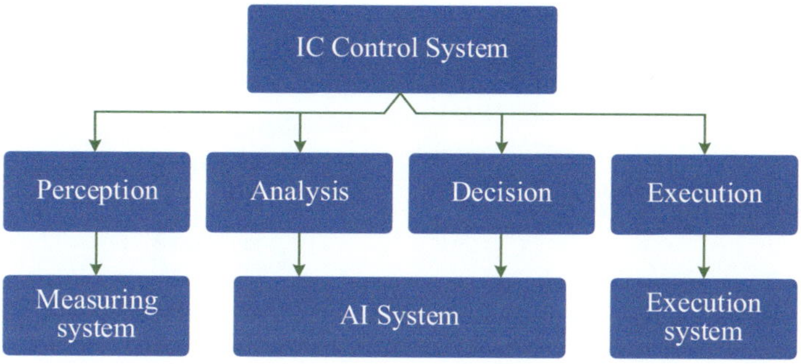

Fig. 1.18 The intelligent compaction control system is divided into three subsystems

between the roller and the filler body and performing various signal processing to obtain data reflecting the mechanical properties of the filler body; the artificial intelligence system located in the controller is responsible for analyzing the data (including data describing the fill properties and compaction process data), and making decisions on how to compact; the execution system (also known as the action system) carries out compaction operations according to the decision instructions, realizing feedback control of the compaction process and fill.

Regarding the AI system (i.e., the intelligent model), it needs to learn and master skills in advance to function effectively. Based on experience, you can choose intelligent models, such as expert systems, neural networks, and reinforcement learning (Fig. 1.17), to complete learning and work tasks independently or in combination.

(4) **Basic characteristics of intelligent compaction**

According to the basic characteristics of intelligent construction, intelligent compaction should have the autonomous ability of "perception, analysis, decision-making, execution." The following is a brief explanation.

1) **Perception ability**

The ability to perceive is key to obtaining compaction data. During the compaction process, the measurement system measures the dynamic response of the vibrating drum (analog signal) through sensors, then converts it into a digital signal (digitization, a misused professional term), and finally performs further processing through the calculation module (microprocessor) to obtain compaction quality information (such as modulus), which is provided to the AI system for analysis and decision-making. Figure 1.19 illustrates the perception process, which is an extension of Fig. 1.17.

In engineering construction, perception is the main technical means to obtain various information (data), which is not only related to electronic devices such as sensors and data collectors but also closely related to relevant professional knowledge (involving perception methods), such as the core module in the above figure, which requires a lot of knowledge support behind it.

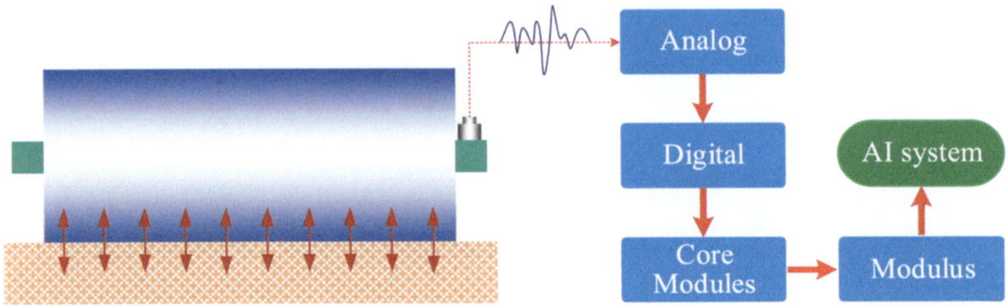

Fig. 1.19 Perceiving compaction quality information

2) **Analytical ability**

The controller (embedded AI system) analyzes various data (such as compaction degree, compaction stability, and compaction uniformity) based on the perceived compaction quality information and the compaction control criteria, providing a basis for decision-making.

Note that before the AI system has analytical capabilities, it needs to undergo learning (training) and acquire enough "knowledge" to make correct analyses and decisions based on the information.

3) **Decision-making ability**

Decision-making primarily determines whether to continue compaction, how to compact (i.e., how to adjust the compaction process), and issues execution instructions based on the analysis results. Analysis and decision-making complement each other: analysis is the cause and decision-making is the result. Similarly, the AI system must first learn about decision-making.

4) **Execution ability**

For intelligent compaction, execution is a compaction operation and involves feedback control. Based on the decision results, the control system issues instructions for feedback control measures during the fill and compaction process, instructs workers to improve the fill, and directs the roller to perform compaction operations to achieve better compaction quality.

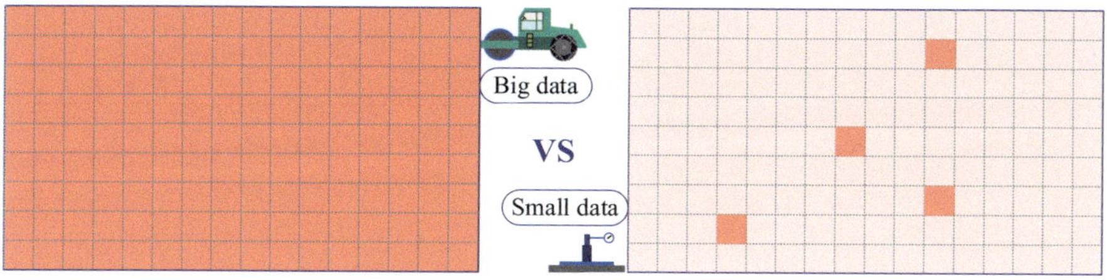

Fig. 1.20 Big data and small data

(5) **Data and value**

Intelligent compaction enables continuous detection of the entire compaction process across the entire compaction surface. The data obtained (such as modulus) constitute the entirety of the compaction surface (which can be understood as the population in statistics), which is a kind of "big data." In contrast, the data obtained from the plate load test is a sample belonging to the category of "small data," as shown in Fig. 1.20. They each have different usage values and can also be used in combination. See Sect. 3.7 for details.

It should be noted that continuous detection and point detection are not issues of data volume but issues of the relationship between the population and the sample. Since intelligent compaction generates large amount of data, its compaction its compaction quality evaluation and control system will naturally evolve, enabling various analyses. From this, valuable information and knowledge can be mined, and more results can be obtained. Additionally, it is also the initial state collection of highways and railways during the maintenance period. A detailed introduction will follow later.

As a pioneer in the intelligent construction of transportation infrastructure, intelligent compaction has led the way in implementing the "perception, analysis, decision-making, execution" approach, serving as an exemplary role in promoting its development. Next, a comprehensive introduction to intelligent compaction control technology will begin.

Chapter 2
Theoretical Basis

Intelligent compaction is the product of multidisciplinary integration. To endow it with the basic abilities of "perception, analysis, decision-making, execution," it requires support from mechanics, automatic testing, AI, control, and professional knowledge. Familiarity with and staying up-to-date on the relevant theories and developments in intelligent compaction are helpful for a better understanding and application of this technology and are also essential for in-depth research.

2.1 Understanding the Compaction Process

Intelligent compaction is a feedback control implemented in the compaction process. The complexity of the compaction process greatly affects the measurement (perception) results of intelligent compaction and determines the difficulty of model establishment. Familiarity with the compaction process is a necessary foundation for establishing models and implementing feedback control.

Compaction is a complex interaction between the roller and the fill, mainly manifested in two aspects. Firstly, the action mode of the roller changes dynamically. Secondly, the deformation of the fill is a mixture of elastic and plastic deformation, as shown in Fig. 2.1.

The complexity of the compaction process increases the difficulty of establishing the theory of intelligent compaction and affects "perception, analysis, decision-making, and control." It is one of the difficulties of intelligent compaction technology.

(1) **Mode of action of the roller**

The roller's mode of action is divided into static compaction, impact compaction, and vibration compaction. Vibration compaction is further divided into circular vibration

G. Xu et al., *Pioneer of Intelligent Construction - Intelligent Compaction*,
https://doi.org/10.1007/978-981-95-4184-3_2

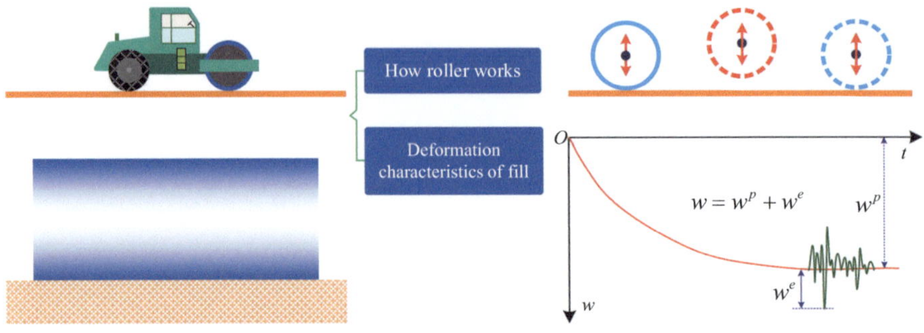

Fig. 2.1 Representation of the complexity of the compaction process

(ordinary vibration roller), directional vibration (vertical vibration roller), and horizontal vibration (oscillating roller), each of which is suitable for different types of fill compaction. Here, the action characteristics of the ordinary vibration roller are detailed.

1) **Vibration characteristics of the vibrating drum**

The vibration action of the vibrating roller is generated by the vibration shaft in the vibrating drum (steel drum), which drives the eccentric block to rotate. If the mass of the eccentric block is m, the eccentricity is e, and the rotation speed is n, then the excitation force generated is $P\sin\omega t$, its amplitude is $P = me\omega^2$, and the vibration frequency is $f = \omega/2\pi = n/60$. The three elements that determine the size of the excitation force are m, e, and n (ω). If you want to change the size of the excitation force, you can achieve it by adjusting any of these elements, or you can adjust these three elements in combination. If there are multiple sets of eccentric blocks with different directions, the excitation force will be their resultant force projected in the vertical direction. Figure 2.2 is a schematic diagram of the generation mechanism and the change in the rule of the excitation force of the vibrating roller.

The excitation force changes according to the sine curve rule, which is called simple harmonic motion in vibration theory. If the excitation force is regarded as an input of the steel drum (system), then the steel drum will produce a response—displacement, speed, and acceleration. In the case of the steel drum being suspended,

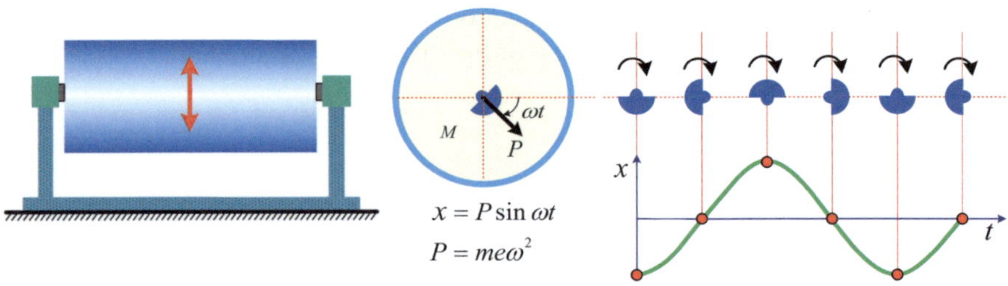

Fig. 2.2 Generation mechanism and change rule of excitation force

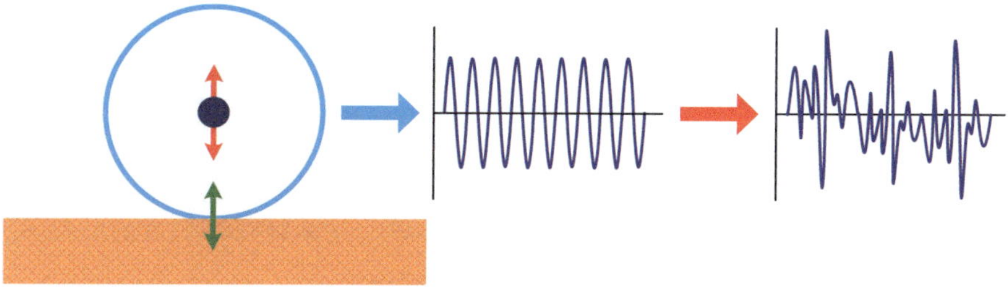

Fig. 2.3 Changes in the vibration form of the steel drum during the compaction process

the response rules of the steel drum's displacement, speed, and acceleration are also sine curves, and the vibration frequency is also ω. Still, the response is different from the excitation force by a phase angle (which changes with the compaction state).

During compaction, a strong interaction occurs between the steel drum and the fill, resulting in a reaction (resistance) from the fill that serves as another input to the steel drum. Under the joint action of the excitation force and the resistance, the vibration form of the steel drum will change greatly. Not only will the amplitude change, but the frequency component will also change, and the phase difference will also change, as shown in Fig. 2.3.

In fact, the interaction between the roller and the fill exhibits complex, nonlinear dynamic behavior, which makes it difficult to establish a compaction model. When this nonlinear action is very strong, it changes the roller's mode of action, producing an extreme phenomenon: the roller bounces, which has a greater impact on establishing the compaction model.

2) **Two types of compaction**

Vibratory rollers employ two types of compaction methods during the compaction process: rolling contact and bouncing (see Fig. 2.1). Under normal circumstances, the interaction between the roller and the filler is closely coupled. That is, the steel drum and the fill are coupled, which is a common compaction method, and the mechanical model is generally established based on this coupling. Although the vibration mode of the steel drum will also change, it remains within a controllable range, and solutions are available.

As the filler's stiffness increases, the resistance encountered by the roller during compaction will also increase. When it reaches a certain level, the steel drum will bounce, losing the coupling effect with the fill. The direct consequence is that it is easy to crush the surface fill. However, a more serious consequence is that the difficulty of establishing a mechanical model will further increase, and all coupled models will no longer be applicable. Although it is possible to consider using a collision (impact) model, it will be more complex.

To be realistic, the bouncing phenomenon often occurs when the roller is compacting, due to a mismatch between the roller and the fill. If it is an intelligent roller, it will control the occurrence of this phenomenon by adjusting the frequency and amplitude.

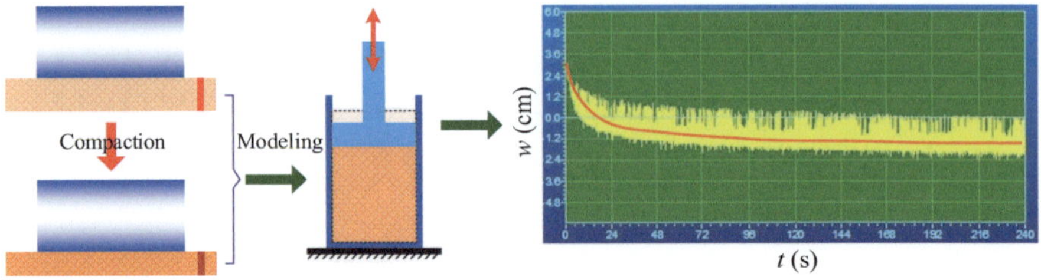

Fig. 2.4 Vibration compaction deformation characteristics of the fill

In addition, during the compaction process, another phenomenon will occur, such as the vibration frequency (ω) fluctuating. This phenomenon is common in early vibratory rollers. If the vibration frequency fluctuates greatly, it will cause greater fluctuations in the exciting force ($P = me\omega^2$), which will inevitably cause a large fluctuation in the compaction force (a concept different from the exciting force), not only artificially causing uneven compaction, but also affecting the measurement work of intelligent compaction. Therefore, the vibration frequency needs to be controlled within the specified fluctuation range.

(2) **Vibration compaction deformation characteristics of the fill**

Compaction is the deformation of the fill during compaction. Under the action of the vibratory roller, the fill gradually changes from its initial loose state to a compact state, undergoing a transition from large plastic deformation to small deformation. Because it is vibratory compaction, there is also the problem of elastic deformation (w^e), that is, $= w^p + w^e$. When the compaction is over, the plastic deformation should be 0, that is, $w^p = 0$. At this time, the fill becomes an elastic body, and the elastic deformation should be a fixed value, that is, $w^e = C$. The compaction deformation curve shown in the right figure of Fig. 2.4 is obtained from indoor vibratory compaction simulation tests, and it is difficult to measure this deformation process on the roller.

The complexity of the compaction process first manifests in the roller's compaction action on the fill, which involves two modes of rolling contact and bouncing. Secondly, the compaction process of the fill involves both plastic deformation and elastic deformation. In actual compaction operations, these manifestations are often mixed, which adds considerable difficulty to establish the model and conduct testing. Choosing the right model to describe this process is crucial. Additionally, when establishing a model, it is necessary to consider whether the information is useful and can be measured accurately.

2.2 The Basis of Perception: Browsing the Mechanical Model in Compaction

According to the basic requirements of intelligent compaction, the first step is to correctly "perceive" how the mechanical properties of the fill are changing. For this, a mechanical model of the compaction problem needs to be established. The model should include parameters that reflect the filler's performance, which can be obtained through testing.

The fill, under the action of the roller, undergoes a change from large plastic deformation to small plastic deformation to elastic deformation. This is a mixed problem of plasticity and elasticity. It remains challenging to fully and accurately simulate and solve this change process, and simplification is necessary. The basic principle of simplification is that it must focus on key points, solve practical problems, and meet engineering accuracy requirements.

Before establishing the compaction model, it is necessary to first develop a basic model to describe the vibratory roller and the fill, which will serve as the basis for the compaction model.

(1) **Basic model**

Intelligent compaction involves two parts: the roller and the fill. The form in which they appear in the compaction model is very critical and determines the difficulty of solving the model.

1) **Model describing the vibratory roller**

In the compaction operation, the main role is played by the roller's vibrating drum (steel drum), and the sensor for continuous measurement in intelligent compaction is also installed on it. Therefore, the so-called vibratory roller model refers to the vibrating drum. The vibrating drum can be described by two models, as shown in Fig. 2.6.

The first model shown in Fig. 2.6 is mainly used for the design of road rollers, treating the vehicle (the outer frame of the steel drum) separately, with the vehicle and the steel drum are connected by springs and damping. When performing vibration simulation analysis of the road roller, the steel drum and the ground are also connected by springs and damping to model the ground's reaction force.

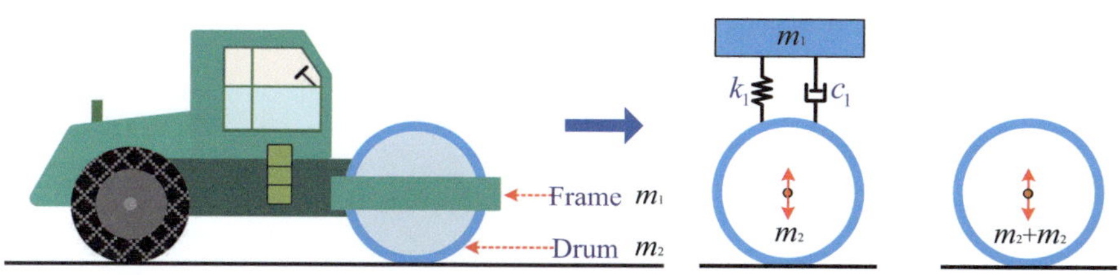

Fig. 2.6 Two models describing the vibrating drum

The second model considers the frame and the vibrating drum together, collectively referred to as the vibrating mass ($m_1 + m_2$). The feature of this model is that it does not consider the details of the road roller but focuses on the interaction with the ground, which can be used in the compaction model of intelligent compaction. The contact between the vibrating drum and the fill can be connected by springs and damping, or the compaction force can represent it.

2) **Model describing the fill**

For the fill, although it has many voids, from a macro perspective-whether it is a highway, railway, airport, or dam, it is considered a continuum (and an elastic body). The advantage of this is that it can be processed using continuous mathematics (calculus).

As mentioned earlier, during compaction, the fill undergoes plastic deformation and elastic deformation. In the case of small deformation, the main difference between the plastic model and the elastic model lies in the stress–strain relationship (constitutive relationship). Still, plastic theory is more complicated and is not widely used in engineering construction. Currently, elastic theory remains dominant, and the theory of elastic small deformation has been widely applied in practical work, as evident in the formulas used in plate load tests, which are based on this theory.

For intelligent compaction, it is currently more convenient to model using elastic theory, that is, considering only the elastic small deformation part. The first model presented in Fig. 2.7 is the elastic half-space model, which incorporates the performance parameters (physical properties) of the fill modulus E, density ρ, and Poisson's ratio μ.

It should be noted that since the fill is compacted in layers, the performance of each layer will not be identical. Therefore, the actual fill is a multi-layer structure; however, for the convenience of handling the problem, it is simplified to an elastic half-space model. The fill parameters obtained by intelligent compaction are not the fill layer itself, but rather the comprehensive value within the roller's depth of influence. Therefore, parameter layering should be carried out. See the later discussion for details.

Another model describing the filler is the discrete model, that is, the second one in Fig. 2.7. This model uses springs k and damping c to simulate the mechanical properties of the fill, with the vibration mass of the soil body concentrated on the mass block. However, determining its size and range is challenging, and in many cases, it is often ignored. The discrete model is often used in vehicle design (including road rollers) and dynamic analysis, and only the reaction force of the fill on the vehicle

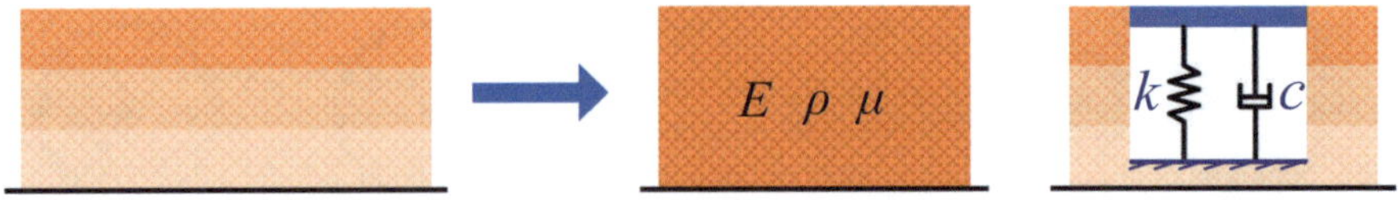

Fig. 2.7 Two simplified models of the fill

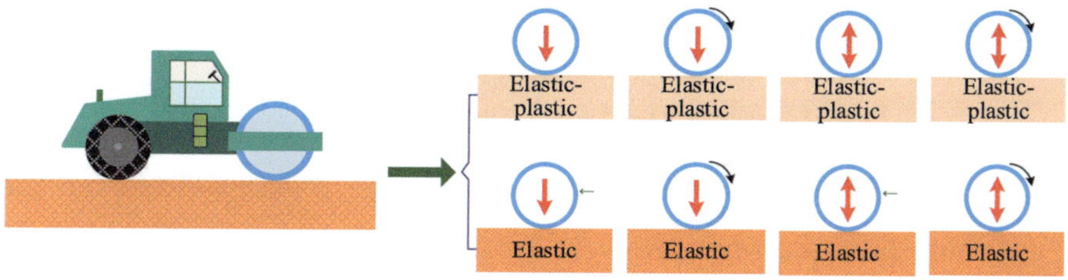

Fig. 2.8 Eight continuum models

needs to be considered. However, for studying the performance of the filler itself, this discrete model has several limitations because it does not account for physical parameters such as modulus, density, and Poisson's ratio, and lacks a concept of thickness, making it impossible to analyze stress and deformation.

The above briefly discusses the models describing the road roller (vibrating drum) and the fill. The compaction model is a certain combination of the road roller model and the fill model. The following introduces several compaction models.

(2) **One of the compaction models—the continuum model**

The continuum model treats the fill as a continuum and combines it with the road roller, which is more consistent with the actual compaction process. Still, it is also the most difficult model. Here, the effect of a single steel drum is considered, and the fill material is not considered in relation to temperature. For asphalt mixtures, the fill is a high-temperature, viscoelastic, and plastic mixture, which is more complicated.

The fill can be divided into two situations: an elastoplastic body and an elastic body. The action mode of the road roller steel drum is divided into four types: static force, static force plus rolling, vibration, and vibration plus rolling. Their combination consists of eight specific models, as shown in Fig. 2.8.

The above eight models basically cover all the models needed for intelligent compaction. The actual action mode of the steel drum only has two types, namely "static force + rolling" and "vibration + rolling," but to simplify the calculation, the rolling problem is not considered at present, and all are regarded as static states. As for the continuum, it currently only considers the elastic state and is modeled as an elastic half-space. Even so, the simplest situation—the static force problem of the steel drum in its static state with the elastic half-space—is still difficult to solve accurately. The focus model of intelligent compaction discussion is "vibration + elastic body," and the answer is also approximate.

The continuum model can be divided into two major categories: static and dynamic. The static model contains only two physical parameters—modulus and Poisson's ratio—lacking density. Although the dynamic model contains these three parameters, it is more difficult to solve. The mechanical model used by intelligent compaction is a combination of these two models.

The continuum model regards the filler body (multi-layer) as an elastic half-space body, and the vibrating drum is regarded as a rigid cylinder. In mechanics, this is the classic problem of the interaction between a cylinder and an elastic half-space, as shown in Fig. 2.9. There is currently no perfect analytical solution.

This model requires that the vibrating drum and the filler body be in close contact and that the two be connected by the action force (compaction force and resistance). If the roller jumps, the model will cease to exist, and a new model must be established.

(3) **Compaction model two—discrete model**

Although the continuum model can simulate the performance of the filler body, its mechanical model is a set of partial differential equations, that are very difficult to solve. To reduce the difficulty of solving, a combination of the discrete model of the filler body and the vibrating drum model can be used, as shown in Fig. 2.10.

In vibration theory, the discrete model is commonly referred to as the "mass-spring-damping" model. Figure 2.10 illustrates two types of discrete body models: the single degree of freedom model and the double degree of freedom model. This type of discrete model is described by ordinary differential equations, which are relatively easy to solve and have a wide range of applications. In compaction quality control, a single-degree-of-freedom model can be used.

In the two models in Fig. 2.10, the vibration mass of the filler body is omitted. Because it is very difficult to determine the filler body's participating vibration mass, it is not the most critical influencing factor for quality control; therefore, there is no need to grasp the main contradiction comprehensively.

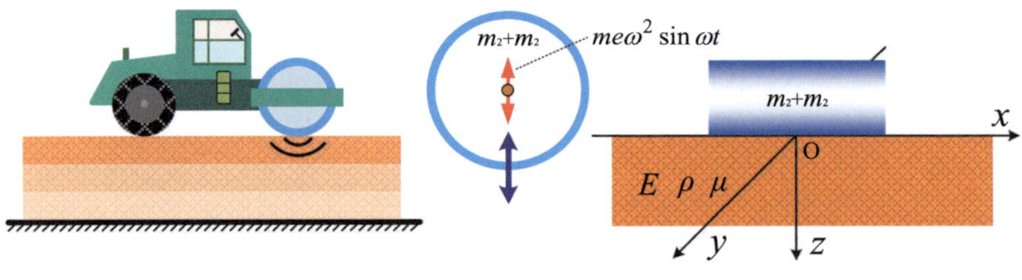

Fig. 2.9 Interaction model of vibrating drum and elastic half-space body

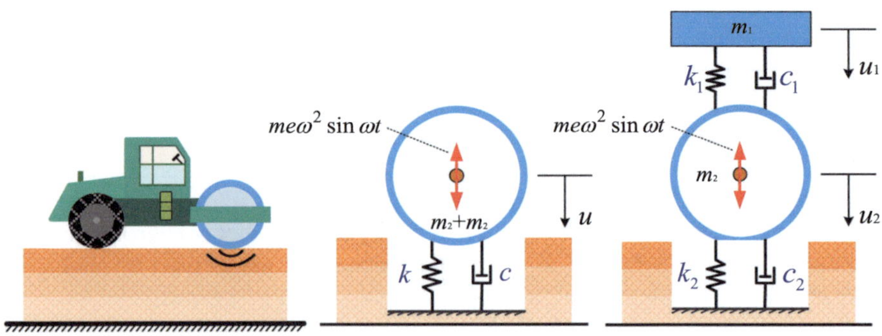

Fig. 2.10 Interaction model of vibrating drum and discrete body

After simplifying the filler body into a spring and damping, the connection between the vibrating drum and the filler body becomes the connection between the vibrating drum and the spring and damping (still part of the contact model), which is tightly coupled over a certain vibration intensity range. But beyond this intensity range, the roller will bounce, and the discrete model will fail.

The "spring-damping" model uses springs and damping to simulate the mechanical properties of the filler body. The size of the spring coefficient (stiffness coefficient) reflects the ability of the filler body to resist elastic deformation. It has a certain correspondence with the modulus in the continuum model. Damping simulates the viscosity of the filler body (such as the cohesion of viscous soil). The discrete model simulates the filler body's ability to resist deformation on a macro scale and is often used in vehicle design and mechanical analysis; however, it cannot perform stress and strain analysis within the filler structure.

The practice has shown that when controlling compaction quality, using a discrete body model is feasible, allowing the spring coefficient k to characterize changes in the compaction state. Additionally, the spring coefficient k exhibits a certain correspondence with the modulus *E*.

——*When the filler body uses different models, the concepts of modulus and spring coefficient will appear respectively, and both can reflect the compaction quality information. So, is there a connection between these two parameters? The modulus parameter appears in the continuum model, reflecting the stiffness of the object (filler body), which is a physical parameter. The properties of the object itself determine the size of the modulus, which remains constant regardless of changes in external force. Still, they can be reflected in the ratio of stress to strain (also the measurement method). The spring coefficient appears in the discrete body model, which also reflects the concept of stiffness, but it does not mean that there is really a spring in the filler body. It is hypothetical and simulated, but it is just used to simulate the elastic properties of the filler body.*

Although these two parameters come from different models for the same object, according to the principle of force and deformation equivalence, using the continuous model and the discrete model, respectively, they can still find similarities between them. Figure 2.11a *shows the relationship between the modulus and the spring coefficient when the rod and the spring reach the same deformation under the same external force, where A is the cross-sectional area of the rod.*

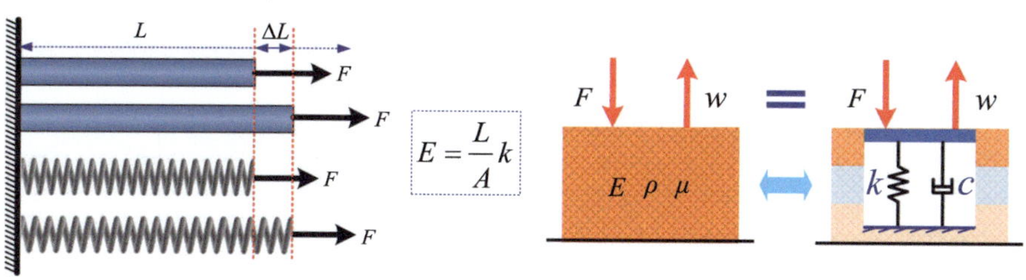

(a) Rod and spring *(b) Continuum and discrete body*

Fig. 2.11 Correspondence of different models

It can be observed that, from the perspective of force and deformation, it is feasible to use a spring to simulate a continuum. This is why, when analyzing the upper structure and vehicles, the structure of road and railway engineering is generally simplified to a spring combination model.

Similarly, in Fig. 2.11b, *the same external force is applied to the continuum and the discrete body, respectively. If the resulting displacement is the same, it is said that these two models are equivalent, from which the relationship between the modulus and the spring constant can be inferred.*

(4) Solution to the problem of road roller bouncing

Whether the model is continuous or discrete, the road roller and the filler body must be in close contact. That is, they are coupled together. This is not only the need for the mechanical model itself, but also the requirement of the test.

The requirement for the road roller and the filler body to be in close contact implies that the displacement of the vibrating drum is coordinated with the deformation of the filler body surface. Intelligent compaction uses this coordination relationship to indirectly obtain the deformation information of the filler body surface (compacted surfaces) by continuously measuring the dynamic response on the vibrating drum, thereby identifying the modulus. If the road roller bounces, this coordination relationship will no longer hold, and the compaction model will need to be reconsidered.

Since the road roller and the filler body are not fixed together, bouncing sometimes is unavoidable. The main reason is that the filler body is too hard (the stiffness is too large). That is, the resistance is too large. The steel drum and the compaction surface have changed from close contact to collision, as shown in Fig. 2.12. At this time, the deformation of the compaction surface cannot be reflected in the steel drum's displacement, and other methods need to be considered.

During the collision process (also called the impact process), the steel drum starts to contact the compaction surface at a very high speed (v), and the speed will gradually become 0 with the compression deformation, reaching the maximum deformation, and then start to rebound until it separates from the compaction surface. The greater the stiffness, the shorter the contact time. Theoretical analysis shows that the contact time (t) is closely related to the mass of the steel drum, the drum width, and the stiffness of the filler body. A collision model can be established to simulate this process.

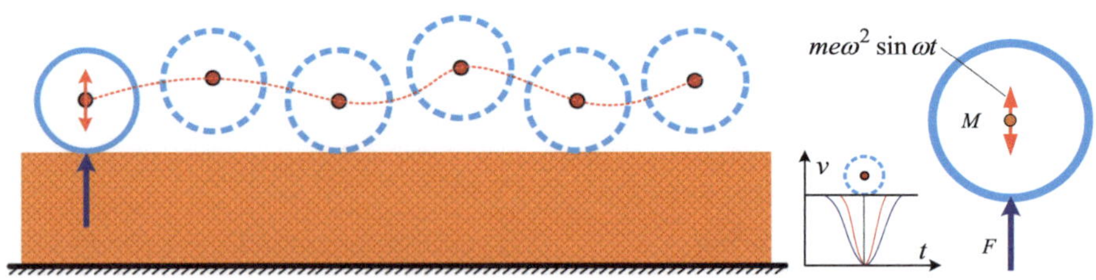

Fig. 2.12 Collision problem during compaction

In the case of a certain steel drum, the contact time is mainly related to the stiffness of the filler body (Fig. 2.12). According to actual observations, as long as the modulus of the filler body is large enough, the road roller will bounce. According to the collision model, the modulus of the filler body can be determined by extracting the contact time using specific techniques.

Another approach to addressing the bouncing problem is to estimate the deformation of the compaction surface based on the displacement of the steel drum. The general rule is that the smaller the deformation of the compaction surface, the larger the displacement of the steel drum bounce. If the deformation of the compaction surface can be estimated, the modulus can be determined based on the continuum model (refer to as the impact modulus).

The control indicators of compaction quality are modulus and density. Regardless of the method used, the control indicator problem should be addressed first. The various compaction models introduced earlier aim to select appropriate control indicators. Due to the complexity of the problem, certain simplifications were made when establishing various compaction models. Still, it is necessary to meet the requirements of engineering accuracy and be able to perceive available information through testing technology.

For intelligent compaction, selecting control indicators is a primary task in the perception stage. Intelligent compaction is essentially a dynamic problem of the interaction and feedback control between the road roller and the filler body. The control indicator used should be a mechanical parameter. The modulus is, of course, the best choice. The corresponding model must be a continuum model and should be a dynamic model.

2.3 Dynamic Model

Based on the previous analysis, the description of the vibration compaction problem can be carried out according to the following scheme: the multi-layer filler body is equivalent to an elastic half-space body, considering the vibration characteristics of the steel drum, and establishing the model using dynamic methods, as shown in Fig. 2.13. If the acceleration of the vibrating drum is 0, it degenerates into a static problem.

The theory of an elastic half-space is well established, and the main differences across specific problems lie in the boundary conditions. The primary challenge of the compaction problem is addressing the contact issue between the steel drum and the elastic body. The distribution of contact stress is shown in Fig. 2.13, divided into static and dynamic types. Currently, it is primarily static; however, when calculating the contact force, a dynamic method based on a single-degree-of-freedom model is employed.

The continuum model employs a set of partial differential equations, comprising balance equations, geometric equations, compatibility equations, and constitutive

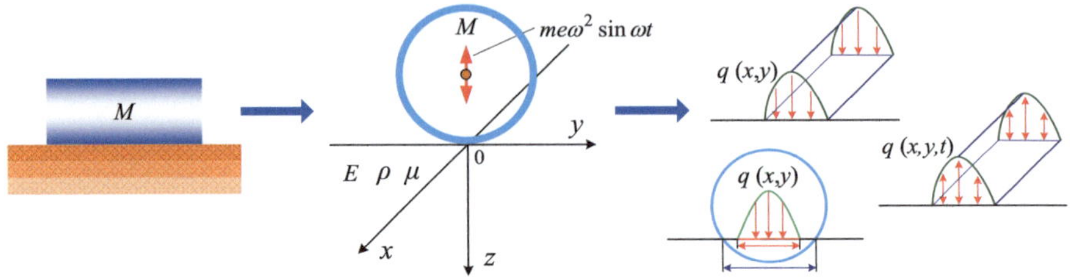

Fig. 2.13 Dynamic model of intelligent compaction

equations. According to the theory of an elastic half-space, the expressions of a point in the elastic body can be directly written out.

(1) **Balance equation (motion equation)**

The balance equation describes the balance state of a point in the elastic body after being subjected to a force, which is derived from Newton's second law. Since the position of a point in the elastic body needs to be represented by three coordinates, it is a multivariate function. It needs to be in the form of partial differentiation.

$$\begin{cases} \dfrac{\partial \sigma_x}{\partial x} + \dfrac{\partial \tau_{yx}}{\partial y} + \dfrac{\partial \tau_{zx}}{\partial z} = \rho \dfrac{\partial^2 u}{\partial t^2} \\ \dfrac{\partial \tau_{xy}}{\partial x} + \dfrac{\partial \sigma_y}{\partial y} + \dfrac{\partial \tau_{zy}}{\partial z} = \rho \dfrac{\partial^2 v}{\partial t^2} \\ \dfrac{\partial \tau_{xz}}{\partial x} + \dfrac{\partial \tau_{yz}}{\partial y} + \dfrac{\partial \sigma_z}{\partial z} = \rho \dfrac{\partial^2 w}{\partial t^2} \end{cases} \tag{2.1}$$

In the formula, u, v, and w are the displacement components in the direction of x, y, and z, respectively, and ρ is the density.

(2) **Geometric equation**

The geometric equation primarily uses the object's displacement and strain to describe its motion or deformation. Under the action of external forces, each point inside the object undergoes displacement, which is defined as a change in position. The change in shape is called strain. Linear strain (positive strain) is represented by ε, and shear strain (angular strain) is represented by γ. Their relationship it as follows.

$$\begin{gathered} \varepsilon_x = \frac{\partial u}{\partial x}, \quad \varepsilon_y = \frac{\partial v}{\partial y}, \quad \varepsilon_z = \frac{\partial w}{\partial z} \\ \gamma_{xy} = \frac{\partial v}{\partial x} + \frac{\partial u}{\partial y}, \quad \gamma_{yz} = \frac{\partial w}{\partial y} + \frac{\partial v}{\partial z}, \quad \gamma_{zx} = \frac{\partial u}{\partial z} + \frac{\partial w}{\partial x} \end{gathered} \tag{2.2}$$

(3) **Strain compatibility equation (compatibility equation)**

The above six strain components are expressed through three displacement components, indicating that there must be an inherent connection among them to ensure the continuity of the object. This relationship is known as the strain compatibility equation, also called the compatibility equation. The expression is as follows.

$$\begin{aligned} \frac{\partial^2 \varepsilon_y}{\partial x^2}+\frac{\partial^2 \varepsilon_x}{\partial y^2}=\frac{\partial^2 \gamma_{xy}}{\partial x \partial y} \quad \frac{\partial}{\partial x}\left(\frac{\partial \gamma_{zx}}{\partial y}+\frac{\partial \gamma_{xy}}{\partial z}-\frac{\partial \gamma_{yz}}{\partial x}\right)=2\frac{\partial^2 \varepsilon_x}{\partial y \partial z} \\ \frac{\partial^2 \varepsilon_z}{\partial y^2}+\frac{\partial^2 \varepsilon_y}{\partial z^2}=\frac{\partial^2 \gamma_{yz}}{\partial y \partial z} \quad \frac{\partial}{\partial y}\left(\frac{\partial \gamma_{xy}}{\partial z}+\frac{\partial \gamma_{yz}}{\partial x}-\frac{\partial \gamma_{zx}}{\partial y}\right)=2\frac{\partial^2 \varepsilon_y}{\partial z \partial x} \\ \frac{\partial^2 \varepsilon_x}{\partial z^2}+\frac{\partial^2 \varepsilon_z}{\partial x^2}=\frac{\partial^2 \gamma_{zx}}{\partial z \partial x} \quad \frac{\partial}{\partial z}\left(\frac{\partial \gamma_{yz}}{\partial x}+\frac{\partial \gamma_{zx}}{\partial y}-\frac{\partial \gamma_{xy}}{\partial z}\right)=2\frac{\partial^2 \varepsilon_z}{\partial x \partial y} \end{aligned} \tag{2.3}$$

(4) **Constitutive equation (stress–strain relationship)**

The above equations are applicable to any continuous medium; only the constitutive equation accounts for the medium's characteristics, expressing the relationship between stress and strain. The constitutive equations of different media are generally distinct (although the equilibrium and geometric equations remain the same), which is the primary feature distinguishing them. For isotropic linear elastic continuous media, the following constitutive equation can be obtained according to the generalized Hooke's law.

$$\begin{aligned} \varepsilon_x &= \frac{1}{E}[\sigma_x-\mu(\sigma_y+\sigma_z)], \quad \gamma_{xy}=\frac{2(1+\mu)}{E}\tau_{xy} \\ \varepsilon_y &= \frac{1}{E}[\sigma_y-\mu(\sigma_z+\sigma_x)], \quad \gamma_{yz}=\frac{2(1+\mu)}{E}\tau_{yz} \\ \varepsilon_z &= \frac{1}{E}[\sigma_z-\mu(\sigma_x+\sigma_y)], \quad \gamma_{zx}=\frac{2(1+\mu)}{E}\tau_{zx} \end{aligned} \tag{2.4}$$

where E is the elastic modulus of the medium, and μ is the Poisson's ratio of the medium.

Modulus, density, and Poisson's ratio reflect the characteristics of the medium itself, are characteristic parameters, and reflect the physical properties, which is why they are called physical parameters. After establishing the above equations, various mathematical tools can be used to solve them. Practice has shown that the difficulty of elastic theory lies not in establishing model equations, but in solving them.

The derivation and understanding of the basic equations of elasticity are not complicated—symbols confuse many people. The brief explanation here is intended to help readers understand the essence and clarify basic concepts.

The stress and strain at any point in the elastic body are related to the position and are functions of three-dimensional coordinates. Since there is no specific expression, it can only be approximated by the Taylor formula, so familiarity with the Taylor formula is key. In Fig. 2.14, *the function without a specific expression $f(x)$, at x_0, can be represented by the approximate form of the Taylor series. The figure shows a first-order approximation.*

If it is a multivariable function (x, y, z), it can also be represented similarly at (x_0, y_0, z_0), as shown in Fig. 2.14. *The derivative of a multivariable function uses the partial derivative symbol, which is why the partial derivative symbol appears in the previous formula. For an*

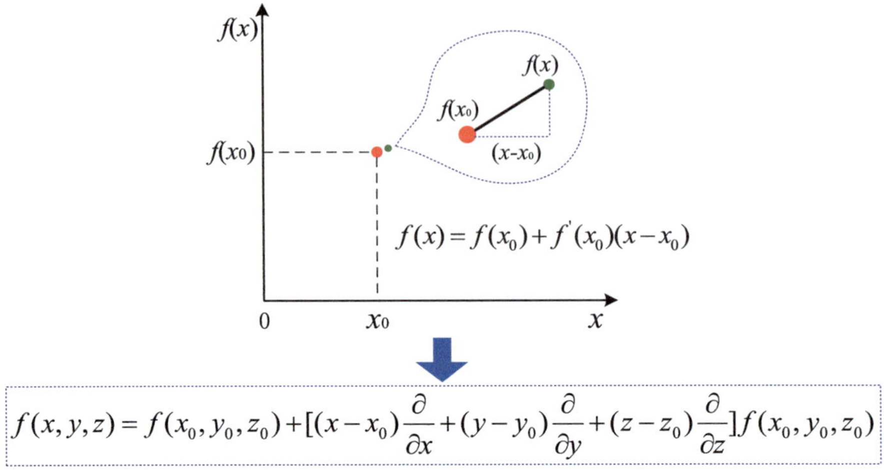

Fig. 2.14 Illustration of the Taylor series expansion of the function near point x_0

unknown function, using a first-order approximation of the Taylor formula can generally meet the accuracy requirements and has been widely used in many fields. It is a tool for approximating nonlinearity to linearity and transforming functions without specific expressions into more expressive forms. The following gives the expression of stress at a point in the elastic body.

After the continuous medium is subjected to a force, additional stress will occur in the body. The stress state at a certain point can be characterized by the stress components on each face of the small cube (microelement), as shown at point M in Fig. 2.15.

The stress at point N near point M can be expressed according to the Taylor formula of the multivariable function (readers can derive it specifically). Since it is a small deformation, it can meet the requirements by approximating to the first order in engineering, so the expressions given in the figure are all linear, which is the origin of linear elasticity.

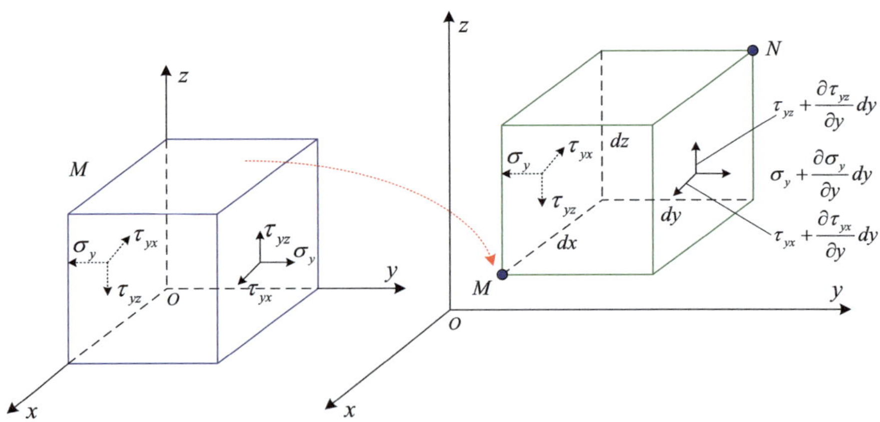

Fig. 2.15 Representation of the stress state at point M and its nearby point N

With the expression of stress, the equilibrium equation can be established according to Newton's law. If you have high school mechanics knowledge, it is actually very simple. Many people are scared by the partial derivative symbol and think it is not very easy. Now, take the y-axis direction as an example. Give the y-axis direction force balance equation, and the rest can be inferred by analogy.

According to mechanical knowledge, there is a force in the y-axis direction on each face, and the size of the force is the product of stress and area, so

$$\left(\sigma_y + \frac{\partial \sigma_y}{\partial y} dy\right) dxdz + \left(\tau_{zx} + \frac{\partial \tau_{xz}}{\partial x} dx\right) dydz + \left(\tau_{zy} + \frac{\partial \tau_{zy}}{\partial z} dz\right) dxdy$$
$$- \sigma_y dxdz - \tau_{zx} dydx - \tau_{zy} dxdy = \rho(dxdydz)\frac{\partial^2 v}{\partial t^2}$$

Simplify to get

$$\frac{\partial \tau_{xy}}{\partial x} + \frac{\partial \sigma_y}{\partial y} + \frac{\partial \tau_{zy}}{\partial z} = \rho \frac{\partial^2 v}{\partial t^2}$$

Similarly, other expressions can be obtained. Although the expressions in elasticity mechanics seem complex, they are not complex at all. If you master the expression method of the Taylor series and add basic mechanical principles, you can establish basic equations.

Establishing the basic equations of elasticity mechanics is not difficult. The difficulty lies in how to solve them, which is closely related to the boundary conditions. For the compaction problem, the boundary conditions are the force of the steel drum on the fill and the resulting deformation. For intelligent compaction, the focus is on solving the partial differential equation group and determining the relationship between the force and deformation.

2.4 The Difficulty of Solving and the Identification of Modulus

Although elastic theory can establish a complete set of basic equations, solving them has always been difficult. This is a mathematical problem: how to solve a set of partial differential equations given the boundary conditions. Currently, only a few situations can be solved analytically, such as the effects of concentrated force and disc load, for which analytical solutions are available. However, for the steel drum's action, the problem becomes very complicated.

(1) **The condition of tight contact between the steel drum and the rolling surface**

The force exerted by the steel drum on the rolling surface is called the compaction force, and it is a pair of action and reaction forces that counteract the fill's resistance. The compaction force is generated by the eccentric block inside the steel drum as it rotates, driving the steel drum to vibrate (but not the excitation force), which is why the steel drum is called a vibrating drum.

When the steel drum is in close contact with the rolling surface, there is such a relationship: the compaction force Q is equal to the resistance F, the displacement

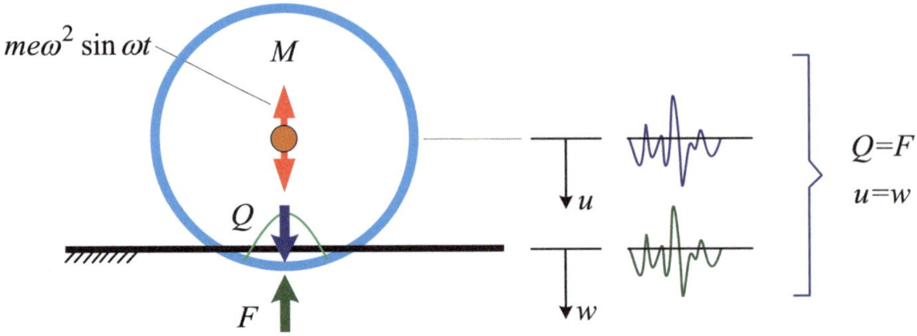

Fig. 2.16 The condition of tight contact between the steel drum and the rolling surface

of the steel drum u is equal to the deformation (displacement) of the rolling surface w, that is, $u = w$, as shown in Fig. 2.16.

The tight contact between the steel drum and the rolling surface means that the force and displacement are coordinated. The displacement of the fill surface can be determined by measuring the displacement of the steel drum ($u = w$). When the steel drum bounces, the displacement of the steel drum and the rolling surface are not the same ($u \neq w$), and the difference is significant. Currently, the displacement of the rolling surface cannot be determined by testing the steel drum.

(2) **The distribution form of contact pressure**

The contact pressure between the steel drum and the rolling surface is shown in Fig. 2.17. The contact pressure area is rectangular. Where L is the length of the steel drum, R is the radius, and $2a$ is the width of the contact pressure, which is generally less than the contact width between the steel drum and the rolling surface.

(3) **The difficulty of solving**

The difficulty of elastic theory is how to solve it according to the given boundary conditions. Currently, only static problems with circular or elliptical contact pressure

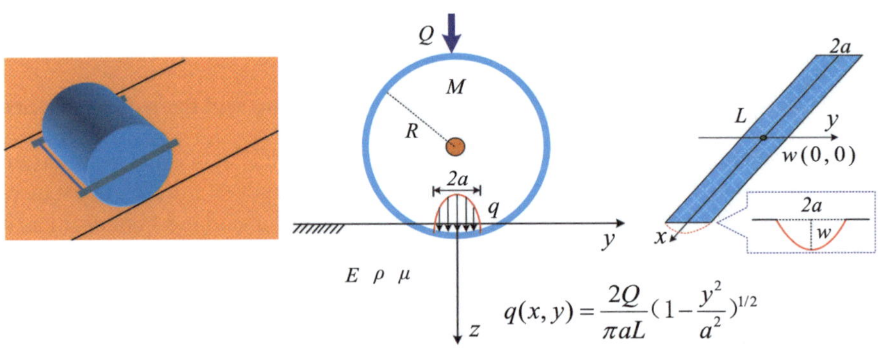

Fig. 2.17 Contact pressure between the steel drum and the rolling surface

areas can be solved (for example, the action area of the bearing plate is circular, allowing for an analytical solution). However, for situations such as the rectangular action area of a steel drum, it isn't easy to obtain an analytical solution. The real difficulty is determining which mathematical tools to use for processing.

In Fig. 2.17, the contact pressure is expressed in the form of (x, y), but there is a prerequisite—the length of the steel drum (L) must be much larger than the contact width ($2a$), that is, $L \gg 2a$, so that the contact area is equivalent to the limit case of elliptical contact. That is, the long axis is very large, allowing for an approximate solution to be obtained.

(3) **Approximate solution and limitations**

The problem of interaction between the steel drum and the fill has not been truly solved to date. All the solutions now are approximate. The displacement solution under the center point of the steel drum action, as given in Fig. 2.18, is also based on various assumptions and approximate processing. The calculation steps in Fig. 2.18 are briefly explained below.

To calculate the displacement of the center point under the action of the steel drum, the following parameters need to be known in advance: the physical parameters of the fill (E, μ), the vibrating drum parameters (M, R, L), and the vibration parameters (m, e, ω), the acceleration response of the vibrating drum. $\ddot{u}$.

The first step is to calculate the compaction force of the steel drum, Q. This can be calculated according to the single degree of freedom in vibration theory, as detailed in reference [3]. Where λ_1 and λ_2 are empirical correction factors related to the fluctuation of vibration frequency and the phase of steel drum response.

The second step is to calculate the contact width $2a$ of the compaction force and the rolling surface. The width of the contact stress is also the width of the stress distribution, which is related to the compaction state of the fill and the size of the steel drum.

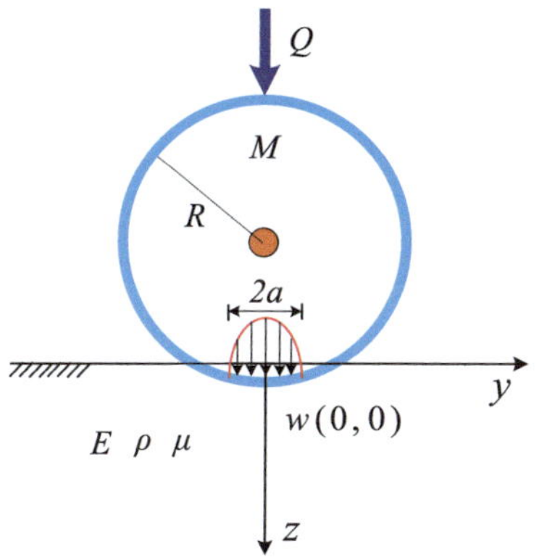

$$(1)\quad Q = Mg + \lambda_1 m e \omega^2 \sin \omega t - \lambda_2 M \ddot{u}$$

$$(2)\quad a = 2\left[\frac{(1-\mu^2)RQ}{\pi L E}\right]^{1/2}$$

$$(3)\quad w(0,0) = \frac{2(1-\mu^2)Q}{\pi L E}\left[\ln\left(\frac{L}{2a}\right) + 1.886\right]$$

Fig. 2.18 Approximate solution of displacement under the center point of the steel drum

The third step is to calculate the displacement. $w(0, 0)$, at the center of the contact area. Although this solution seems to be analytical, it is approximate.

This solution was proposed by Lundberg in 1939. However, this method has significant limitations. The displacement at the end of the steel drum is different from that at the center, which is inconsistent with the actual situation. Therefore, this solution is valid only at the center under the action of the steel drum, and the solutions at other points are flawed. There are many reasons for these flaws, one of which is the unreasonable approximation of the contact pressure distribution form.

In addition, the above solution is static; however, the compaction forces are calculated under vibration conditions and are based on dynamics principles. From the perspective of mechanics theory, this doesn't make much sense. However, in practice, we often see a mixed use of dynamics and statics, and the results are not bad.

(4) **Multiple methods of identifying modulus**

The purpose of solving for the displacement on the surface of the fill is to establish the relationship between compaction force and displacement, thereby identifying the fill's modulus, a key issue in intelligent compaction. There are currently multiple methods for determining the modulus, but few are publicly available. Here are a few ideas for readers to refer to when studying.

The first idea is the direct calculation method. That is, according to the formula shown in Fig. 2.18, directly solving the modulus is the method that many people first think of. Let's try to derive it, and the result is as follows.

$$E = \frac{2(1-\mu^2)Q}{\pi Lw}\left[\ln\left(L/4\left(\frac{(1-\mu^2)RQ}{\pi LE}\right)^{1/2}\right) + 1.886\right] \tag{2.5}$$

If modulus E is regarded as an unknown, it is found that Eq. (2.5) is a logarithmic equation, which is not easily solved; therefore, numerical methods can be used to solve it.

The second idea is the parameter identification method. Parameter identification, also known as parameter estimation, is the "gray box" method in system methods, as shown in Fig. 2.19.

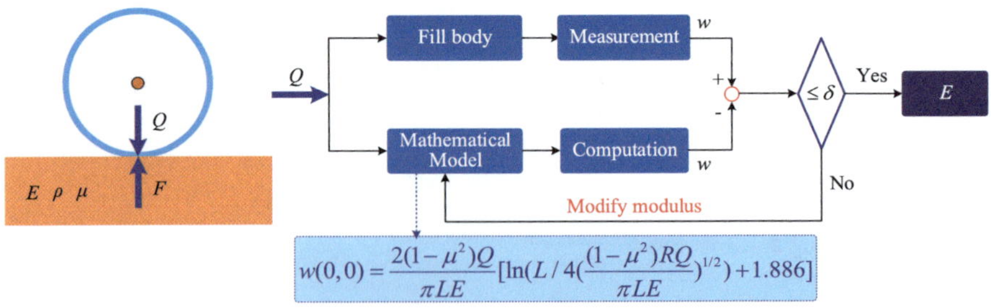

Fig. 2.19 Principle of modulus identification

The parameter identification method first assumes an initial modulus value, substitutes it into the calculation formula, computes the displacement, and then compares the calculated displacement with the measured displacement. If the error is significant, adjust the modulus value accordingly to account for the size of the error. After repeated iterative calculations, the correct modulus value is obtained when the error between the calculated displacement and the measured displacement meets the specified requirements. This method is very similar to the error backpropagation method in ANN (see the introduction of the series). The difference is that there is a fixed model; only the parameters are unknown, and the modulus is the parameter to be adjusted.

The third idea is the experimental method. The basic idea of the experimental method is to study the relationship between the force and deformation of an object, identify the relationship, and then determine the property parameters of the object, specifically the modulus, through experimentation.

A substantial amount of theoretical and practical research indicates a specific relationship between the force and deformation of an object, encompassing the attribute parameter of modulus. Figure 2.20 shows the interaction between rigid bearing plates, flexible bearing plates, rigid spheres, and rigid cones with elastic half-space bodies. After theoretical derivation, the relationship between the force Q and the deformation at the center of the contact area w, which includes the modulus E, can be obtained. Plate load tests are based on the measured force and deformation, and the modulus is determined from this relationship.

Observing the expressions in Fig. 2.20, it can be found that in the relationship between force and deformation, there are three parts: the first part is related to the type and size of the load; the second part is the physical parameters of the elastic half-space body (E and μ); the third part is deformation (displacement). From this, a general expression can be derived, as shown in Fig. 2.21.

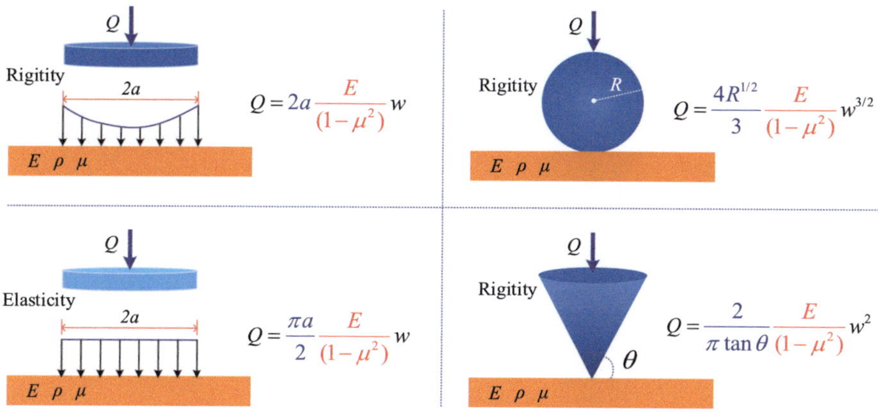

Fig. 2.20 Relationship between four loads and deformation

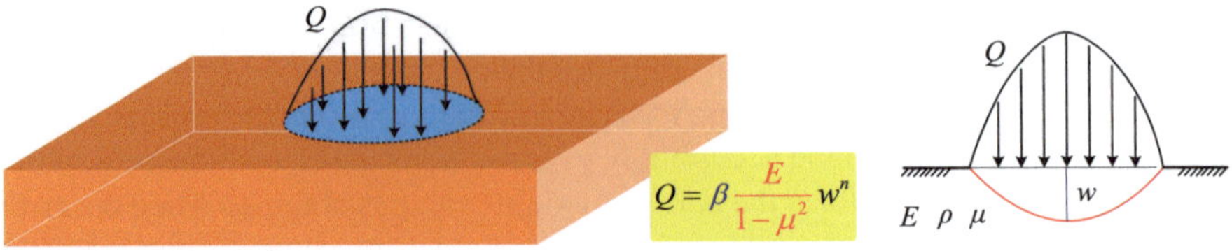

Fig. 2.21 General relationship between force and deformation

In Fig. 2.21, if the surface of the elastic half-space body is subjected to any load F, the displacement at the center of the action area is w, and the following relationship exists.

$$Q=\beta\frac{E}{1-\mu^2}w^n \tag{2.5}$$

In the formula, the distribution of load Q is arbitrary; β is the load shape correction factor, which is related to the distribution shape of the load; n is the exponent of displacement, which is related to the form of load action, for circular load $n = 1$, for spherical load $n = 3/2$.

The above general formula is also applicable to compaction problems. In Fig. 2.18, the interaction between the steel drum and the elastic half-space body can be rewritten as follows.

$$Q=\frac{\pi L}{2}[\ln(\frac{L}{2a})+1.886]^{-1}\frac{E}{1-\mu^2}w \tag{2.6}$$

In the formula, it can also be divided into three parts, with the same meaning as before. The load correction factor β is related to the length of the steel drum and the contact width. Determining β is the key.

Due to the limitations of formula (2.6), we can represent the interaction between the steel drum and the filler body in a more general form.

$$Q=\beta\frac{E}{1-\mu^2}w \Rightarrow \beta=\frac{1-\mu^2}{E}\frac{Q}{w} \tag{2.7}$$

(6) **Relationship between modulus and stiffness coefficient**

In the formula, the load correction factor β must be determined experimentally. If β is determined, the above expression can be used to identify the modulus.

According to formula (2.7), the expression of modulus under different loading methods can be obtained.

$$\frac{E}{1-\mu^2}=\beta_1^{-1}\frac{Q_1}{w_1}=\beta_2^{-1}\frac{Q_2}{w_2} \tag{2.8}$$

In formula (2.8), the left side is the expression of the physical parameters of the continuum. If the filler body is elastic, it can be defined as $K = Q/w$, which represents the concept of stiffness coefficient and the foundation reaction modulus in the Winkler foundation model. The foundation reaction coefficient K_{30} is also defined in this way. By rewriting the formula (2.8) into the following form, the relationship between modulus and stiffness coefficient can be obtained.

$$\frac{E}{1-\mu^2} = \beta_1^{-1} K_1 = \beta_2^{-1} K_2 \tag{2.9}$$

For a specific elastic body, the modulus and Poisson's ratio are fixed physical parameters that and do not change with different loading methods. However, different loading methods have different load correction factors, resulting in varying stiffness coefficients. Therefore, the stiffness coefficient is not an "objective quantity" and is related to the loading method. K and E are very different. In controlling compaction quality, E is a better control index and is more objective. Unless otherwise specified, the continuous control indicators used in this book are all in modulus form.

It should be noted that, although in elastic theory, the modulus does not change with changes in loading method and load, in engineering practice it was found that the modulus also changes with different loading methods and loads. This it caused by the rock and soil's inherent complexity, which has nothing to do with the theory of elasticity. It also shows that elastic theory does not fully and accurately capture the characteristics of the filler body; it is merely an approximate description.

There are several methods for identifying the modulus of the filler body, including simplified formulas and intelligent algorithms. However, all current methods for identifying the modulus are approximate. The modulus obtained based on the response of the roller (E_{Roller}, abbreviated as E_{R}) may be more accurately described as an "estimated modulus." In addition, the modulus obtained by different methods may vary, but they can accurately reflect the characteristics of the filler body and meet the needs of quality control.

Currently, the answers obtained from the vibrating roller are all based on static methods, even though the data used are dynamic, resulting in a quasi-dynamic answer. The mixed solution of dynamics and statics is used here. Although it can be used in engineering, it is lacking and imperfect in mechanics. Therefore, seeking a satisfactory dynamic answer is still the focus of research in this field.

2.5 The Key to Perception: Continuous Testing

The first step in intelligent compaction is to perceive compaction quality information, which requires the help of testing technology. The compaction quality information mentioned here mainly refers to the filler body's performance parameters (modulus). The various methods of identifying the modulus, as discussed earlier, all require the known compaction force of the filler body and the corresponding deformation. However, intelligent compaction directly measures the response signal of the vibrating drum, rather than the compaction force and deformation of the compacted surface, which other methods must determine. Therefore, to obtain the compaction force of the steel drum and the deformation of the compacted surface, the directly measured response must accurately reflect the compaction state; however, this is not easy to achieve.

(1) Continuous testing principle

Intelligent compaction is a continuous test (continuous detection)—continuously measuring the vibration response of the vibrating drum during the compaction process. Its testing principle is not complicated; generally, a microcomputer (single-chip microcomputer) is used to control the collection, conversion, and processing of signals, which are automatically completed, as shown in Fig. 2.22.

In terms of hardware, continuous testing of intelligent compaction is not very different from general vibration testing; the main difference lies in the software. The most distinctive aspect is data processing, which is reflected in two key areas. One is to obtain mechanical performance parameters (such as modulus) of the filler body through the core module; the other is to implement an AI system that enables it to learn, analyze, and make decisions autonomously. This part of the function can also be placed on the information platform and realized through the network for cloud computing.

With the development of technology, new sensors (such as MEMS) have moved the signal conditioning and digitization (A/D) parts into the sensor chip. The signal output from the sensor is a processed digital signal, which is more convenient for subsequent data processing. For more details, see the Perception and Data volume of the series.

(3) Factors affecting continuous testing

Due to the mobility of the compaction operation, the intelligent compaction testing can only be performed on the moving roller. The response signal of the vibrating drum contains two parts of information, as shown in Fig. 2.23. One is the vibration information of the vibrating drum itself, including vibration mass M, excitation force P,

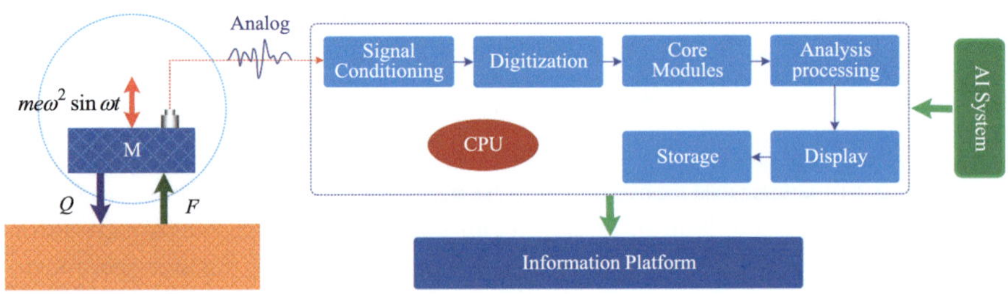

Fig. 2.22 Continuous testing principle

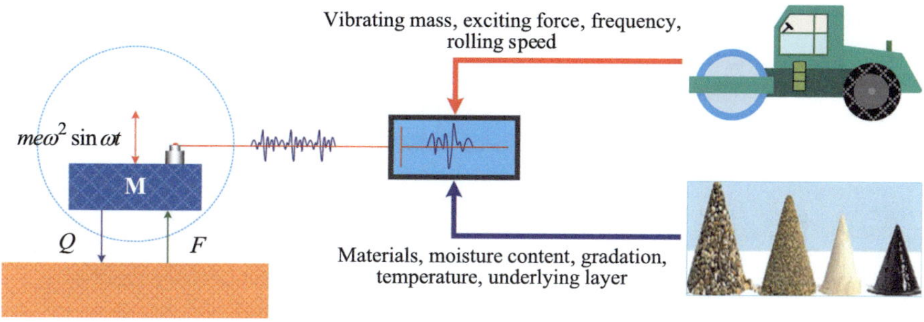

Fig. 2.23 Factors affecting continuous testing

vibration frequency f, and compaction speed v, called vibration compaction process parameters, among which the key parameter is (M, P, f). The second is the information on the filler body, including material type, moisture content, gradation, temperature, and underlying layer. Changes to the information above will affect continuous testing results and are important influencing factors. For intelligent compaction, we hope that test results are more sensitive to materials, moisture content, gradation, temperature, and the underlying layer, which will help better capture changes in compaction state.

The purpose of continuous testing is to detect the compaction state of the filler body. The influence of materials, moisture content, gradation, temperature, and underlying layers on the test results is accurately reflected. What needs to be eliminated is the influence of the loading equipment (vibrating roller), especially the impact of random variations in vibration performance on the test. This requires specific vibration performance requirements for the roller, which are the same as those specified in the loading method for the bearing plate test.

— *—From the perspective of the test, the compaction operation of the vibrating roller is a dynamic test, that closely resembles the plate load test, in which the roller is the loading equipment, and the filler body is the test object. Due to the non-uniformity of the roller type and vibration process parameters and the certain randomness, the steel drum loading is relatively "rough," and the continuous test results are also relatively "rough." The bearing plate test has strict regulations on the loading method. In fact, all mechanical tests have clear requirements for loading equipment and methods to ensure the standardization of the test. As a type of loading equipment, the roller should meet specific requirements for its vibration performance to ensure the reliability of the test results.*

(4) **Vibration performance of the roller**

The practice has shown that different rollers exhibit varying vibration parameters, with their vibration responses also differing. Consequently, the mechanical performance parameters of the obtained filler body are distinct, which is attributed to the characteristics of the filler body (Sect. 1.4). As the loading equipment for continuous

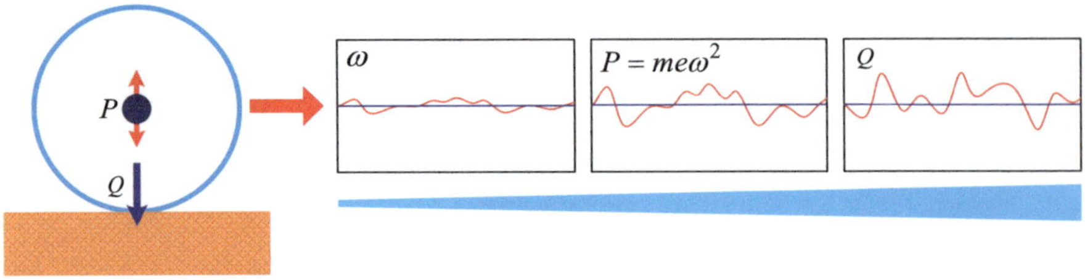

Fig. 2.24 Fluctuation of vibration frequency will cause greater fluctuation of compaction force

testing, the roller should keep the vibration parameters, M, P, ω, constant during the testing process. Otherwise, it will affect the test results. A hidden problem is that the roller's vibration frequency will fluctuate, leading to greater fluctuation in the compaction force Q, as shown in Fig. 2.24.

If the compaction force fluctuates during compaction, it will not only affect test results but also artificially induce uneven compaction. Therefore, the performance of the vibrating roller is specifically stipulated in the Chinese intelligent compaction technical standards (see reference [3]), the purpose of which is to control the stability of the roller's vibration performance. Generally, the fluctuation range of the vibration frequency should not be greater than (0.5–0.6) Hz.

(5) The Influence Range of Road Rollers

The influence range of a road roller refers to the range affected by the vibration of the steel drum in both vertical and horizontal directions. The horizontal influence range is the length of the steel drum, which is relatively fixed. However, the vertical influence range (impact depth) of the road roller varies with the roller's weight, vibration intensity, and fill's compactness, greatly affecting compaction quality and influencing the interpretation of continuous test results.

The purpose of continuous testing is to control the quality of the compacted layer. However, the road roller's impact depth is significantly greater than the thickness of the fill layer. The measured information is the composite value within the impact depth range, which is significantly different from the range used in the plate load test, as shown in Fig. 2.25.

In Fig. 2.25, modulus E_5 of the current compacted layer can be obtained through the plate load test. Still, the test using the road roller yields a composite value across multiple layers (E_1_E_5), which sometimes does not fully reflect the compaction quality of the current layer. Suppose there is a too-hard (or too-soft) interlayer in the underlying layer. In that case, the continuous test results will also increase (or decrease), which can easily lead to a misjudgment of the current layer's quality.

——*The fill is layered (30–40 cm) for spreading and compacting. The task of the road roller is to compact the fill layer into an elastic state that meets the requirements. However, the compaction impact depth of a vibrating road roller is greater than the thickness of the fill layer (e.g., the impact depth of a 16-ton vibrating road roller can reach 1.5 m). The impact*

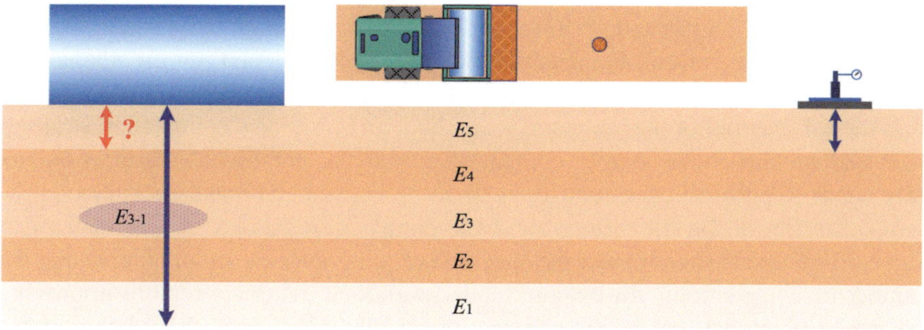

Fig. 2.25 Schematic diagram of the influence range of road rollers and bearing plates

it will have on the current fill layer is worth studying. This will aid in the development of new road rollers and foster the advancement of intelligent compaction technology.

The inconsistency between the test depth of the road roller and the plate load tester can also cause problems when establishing a correlation, and sometimes it is impossible to establish one. To solve this problem, it is necessary to separate the information on the current layer from the comprehensive test results of the road roller. A processing plan will be provided later.

2.6 Tools for Analysis and Decision Making: Intelligent Models

After obtaining compaction quality information (modulus), the next task is to systematically analyze it and decide how to compact it based on the analysis results. With different control methods, the content of analysis and decision-making also varies, as shown in Fig. 2.26.

In point control, manual analysis is employed, typically utilizing a one-vote veto system. If an unqualified point is detected, the entire area is considered unqualified; in continuous compaction control, a computer (machine) is used for automatic analysis

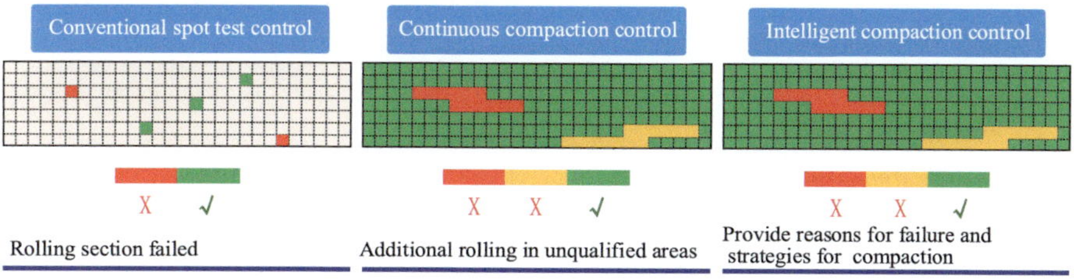

Fig. 2.26 Analysis and decision content of different control methods

to find unqualified areas and perform additional compaction in these areas; in intelligent compaction control, the machine not only needs to point out the unqualified areas, but also needs to analyze the reasons for the failure—whether it is a problem with the fill material or the compaction method, and provide a compaction strategy.

Based on the above simple comparative analysis, intelligent compaction not only needs to perform continuously but also to analyze and make decisions on the compactability of the fill material and the compaction process of the road roller. Some fill materials cannot meet the specified requirements by simply increasing the number of compactions, for example, coarse-grained materials require improving the gradation to achieve better compaction results. How does the machine have this ability? You can choose a suitable intelligent model to achieve this wish. The following introduces several intelligent models that can be used in combination with compaction control, but they are not limited to these, and new methods are still under development.

(1) Expert System

Due to the complexity of the fill material and the diversity of the compaction process, in many cases, the control of compaction quality is not as simple as judging whether the compaction is qualified. There will be many complex situations during the compaction process, such as local inability to compact and filler material being crushed, which require real-time on-site processing. At this time, experienced engineers (including road roller drivers) will be in demand. They can propose treatment suggestions based on specific situations, which is the domain experts' role. If the knowledge of many experts is compiled into the computer, enabling it to handle complex problems, it becomes an expert system.

The expert system originated in 1965 and is the most important method in early artificial intelligence. It is a model of human reasoning ability. This is a computer program that simulates the human brain's reasoning process, enabling the machine to acquire more knowledge. Its core is the knowledge base and inference engine. Its working principle is simple: to find answers in the knowledge base according to the proposed questions. This process is called reasoning, and it can explain why such an answer is chosen.

Although expert systems have numerous shortcomings, they have achieved considerable success in certain professional fields. Even now, it remains the most practical and widely used branch of artificial intelligence.

1) Expression of Knowledge

An expert system is a computer program, and its development process is known as knowledge engineering. The focus of development is the knowledge base and inference engine. The reasoning part has a certain universality, and the most characteristic part is the knowledge base, which is closely related to the application field. A knowledge base consists of several pieces of knowledge and building one involves transforming this knowledge into a form the computer can accept and storing it in a designated location. For compaction control, the key issues are how to acquire compaction knowledge and how to express it effectively.

Knowledge is people's cognition and accumulation of experience about things. The main way for people to acquire knowledge is to learn, to master it through study, practice and books. Traditional expert systems can not learn, and the acquisition of knowledge is done manually, summarizing the experience of experts. For this reason, a specific format needs to be adopted to express knowledge in a way that facilitates easy operation.

In AI, the most commonly used format is the "IF <condition> THEN <conclusion>" form, also known as "rules". Here are a few examples of rules, as shown in Fig. 2.27.

Rules seem simple, but if you want to solve a problem, you need to use many rules to work together. This requires converting various related knowledge into rules, and a significant amount of detailed sorting work is required, necessitating participation from multiple parties and consuming substantial amount of manpower and resources. With the development of intelligent technology, the use of intelligent algorithms, such as neural networks (ANN), to represent rules is on the rise, as described later.

For compaction quality control, converting engineering knowledge related to compaction into corresponding rules is an important task, which is the core of building a compaction knowledge base .

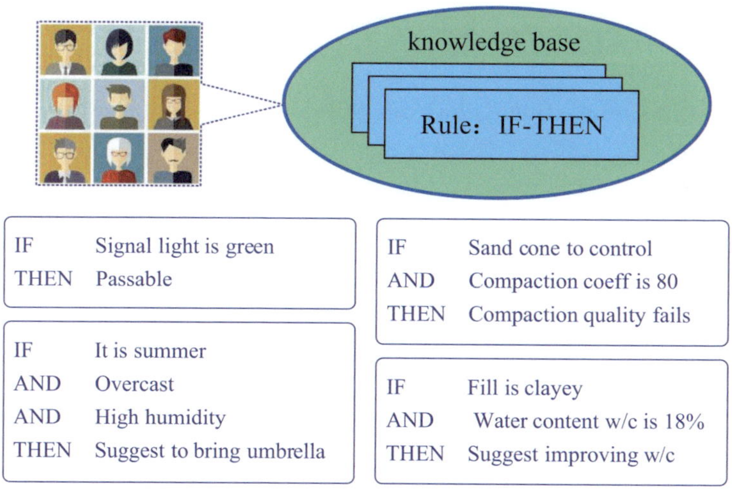

Fig. 2.27 Knowledge transformed into rules

2) **Basic Structure of Expert Systems**

Using rules to represent knowledge is the most common approach in expert systems, and its basic structure is illustrated in Fig. 2.28. The knowledge base is the core, a collection of knowledge that solves problems, stored in the form of rules. Each rule expresses a relationship, a suggestion, an instruction, a strategy, or another concept. When a rule's condition (IF part) is met (true), the behavior after THEN is executed. The database stores a set of facts (initial data for problem-solving, intermediate results, problem-solving results, etc.). The goal is to match the database information with the IF part of the rules in the knowledge base. If the IF part is satisfied, the THEN part (behavior) is executed, and the conclusion to solve the problem is finally obtained.

In Fig. 2.28, the knowledge base contains three rules, and the database contains five facts (information). Based on fact A, the conclusion that Z is true can be obtained. The reasoning process is as follows: based on the given fact A, find the matching rule in the knowledge base, Relu3 is first activated, the conclusion that X is "true" is obtained, then Relu2 is activated, the conclusion that Y is "true" is obtained, and finally Relu1 is activated, the conclusion that Z is "true" is obtained.

There are various forms of reasoning technology in expert systems, each with certain degree of universality. In compaction work, reasoning technology from other fields can serve as reference. Still, the knowledge base must be built separately, considering the characteristics of the professional field.

3) **Building a Compaction Knowledge Base**

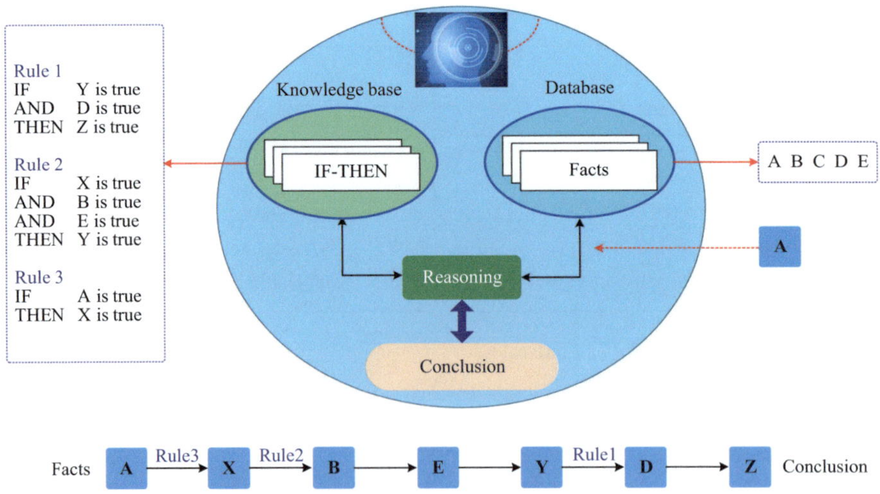

Fig. 2.28 Basic structure and reasoning process of expert systems

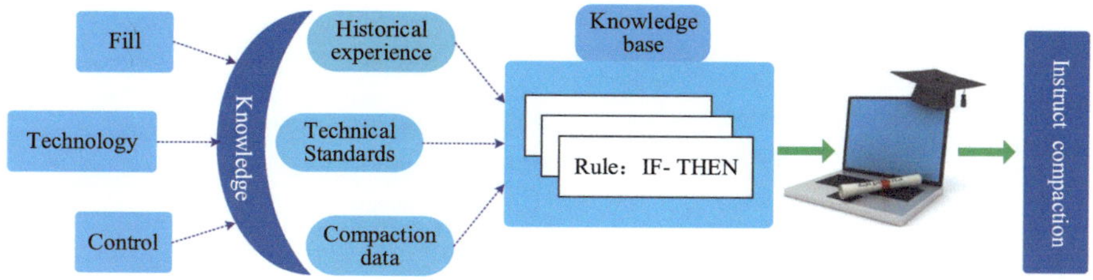

Fig. 2.29 Ways to acquire compaction knowledge

The core of the compaction expert system is the compaction knowledge base. The quality and quantity of knowledge determine the level of the expert system. Knowledge acquisition is the key to building a knowledge base. Figure 2.29 illustrates three methods for acquiring compaction knowledge.

Acquiring compaction knowledge through historical experience mainly refers to the experience of experts and past construction experience, etc.; technical standards are easy to understand, which are regulations for compaction quality control; compaction data refers to the perception data after using intelligent compaction technology, including comparison data of test sections and data obtained from continuous detection, etc. The content of knowledge primarily includes the types and properties of fillers, compaction processes, and control measures, among others. This knowledge needs to be broken down and then converted into several rules in the IF–THEN (also written in lowercase as "if–then") format.

Turning knowledge into rules requires careful consideration and close cooperation among professional engineers (domain experts), knowledge engineers (who design and build expert systems), and users (construction units). This part of the work is time-consuming and laborious, involving issues such as how to extract knowledge from a large amount of information given. In this regard, the mature experience from other fields can serve as reference, and the method of representing knowledge using intelligent models is also under consideration.

(2) **Neural Network**

The Artificial Neural Network (ANN) emerged in the 1980s and is one of the main representatives of intelligent algorithms. Deep learning is one of them. Neural networks are supervised learning algorithms, which require a large amount of labeled data for training (learning) before they can work. Figure 2.30 shows the working and learning process of a simple ANN model (see the introduction of the series for details).

The working process of an ANN is simple. Apart from the input layer, the signal transmission process in the hidden layer (Layer 2) and the output layer (Layer 3) are the same, following the "sum and activation" steps. Taking the unit ① in the hidden layer (Layer 2) in Fig. 2.30 as an example, first, the input of this unit is weighted and summed to get $z_1^{(2)}$, then $z_1^{(2)}$ is put into the activation function $h(z)$ to get the output $y_1^{(2)}$, …, until finally, the output of the ANN $y_1^{(3)}$ is obtained.

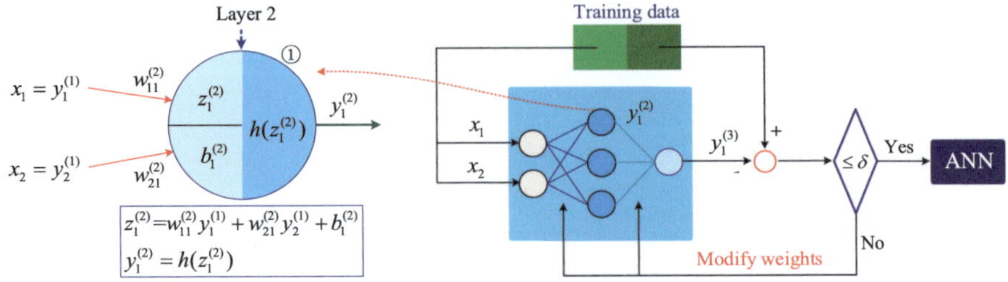

Fig. 2.30 Working and learning process of ANN

The learning process of an ANN is also called training. The training data (data with answers, also known as labeled data) must be split into two parts. One part of the training data is input into the ANN, and the calculation is performed according to the steps outlined in Fig. 2.30. Finally, the output of the ANN $y_1^{(3)}$ is obtained and then compared with the known results (the other part of the training data). If the error exceeds the specified requirement, the weights are modified (see the introduction volume of the series for the modification method), and the calculation is repeated until the error meets the requirement. At this point, the correct ANN model is obtained, and can be used for work.

Including ANN, all supervised learning algorithms currently perform only two tasks: regression and classification. Regression, also known as prediction, is used to infer the inherent regularity of data from existing data to predict new data. Classification is the process of categorizing data according to specific attributes, thereby determining the type of a thing. The first problem in applying intelligent algorithms across industries is how to transform the problems in this field into prediction or classification problems, which require a multi-disciplinary knowledge base.

To be realistic, at present, the application of various intelligent algorithms in engineering construction is not widespread. Most of them make simple predictions. As for the application of classification, there are very few. The primary reason is that it is unclear how to integrate engineering problems with intelligent algorithms. For the analysis and decision-making of intelligent compaction, the key is also how to transform engineering problems into prediction or classification problems for intelligent algorithms.

1) **Prediction**

Prediction is a common problem across various industries, including economic forecasting, structural life prediction, and stock trend prediction. In compaction quality control, there are also many prediction problems, such as predicting point detection results from continuous test results and predicting compaction effects from filling properties and compaction processes. Figure 2.31 provides a simple prediction example, an ANN learns to compare test data and predicts the results of bearing plate tests (such as modulus E_R) based on intelligent compaction detection results (such as estimated modulus E and foundation reaction modulus K_{30}).

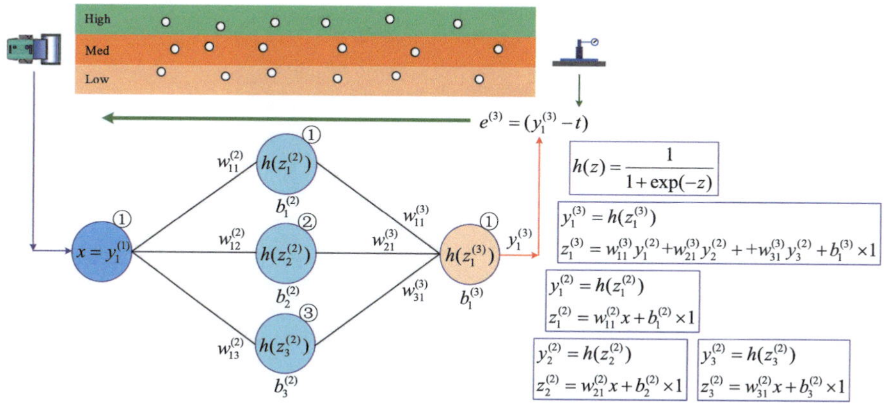

Fig. 2.31 ANN prediction example

The purpose of conducting comparative tests (see the introduction later) is to establish the relationship between intelligent compaction detection results and flat plate load test results. Typically, a linear regression equation is used to represent the relationship between the two variables. Here, an ANN model is used to represent this relationship (but the correlation coefficient cannot be determined). The input layer has one unit (accepting intelligent compaction detection results), the hidden layer has three units, and the output layer has one unit (outputting the prediction results of the flat plate load test). The activation function uses the Sigmoid function.

The learning process of an ANN involves continuously adjusting the weights based on the size of the error between the calculated value and the measured value (t) of the output, until the error meets the specified requirements. The backpropagation error method is generally used to update the weights. For specific operations, please refer to the introduction volume of the series.

The above ANN model is applicable only when the filling material and compaction process are fixed. When the filling material or compaction process changes, the model needs to be retrained to establish a new relationship.

2) **Classification**

Compared with prediction problems, classification problems are more common, such as speech recognition, image (face) recognition, and spam mail recognition, etc., all of which are classification problems, and many problems can also be transformed into classification problems (which need to be explored in practice). If the answer to the classification problem is "yes (true)" or "no (false)," it belongs to binary classification, which can be represented by 1 and 0.

In compaction quality control, compaction quality recognition, compaction process selection, and filling material recognition all belong to classification problems. Figure 2.32 provides an overview of using an ANN for compaction quality classification.

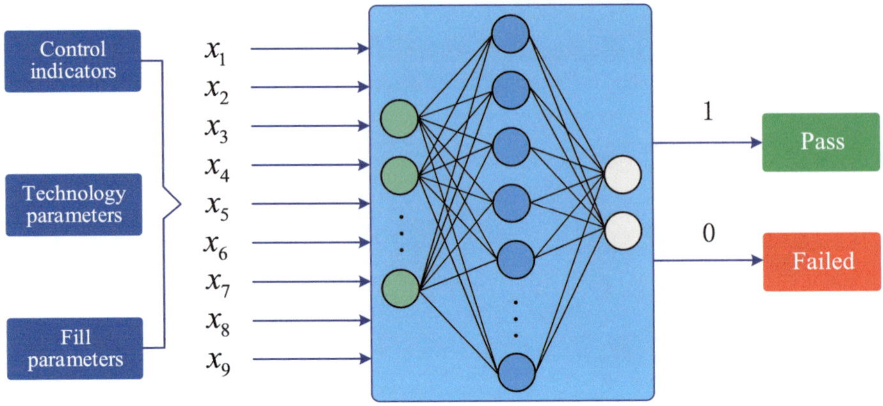

Fig. 2.32 The idea of using ANN for compaction quality classification

According to the compaction standard, compaction quality is judged as qualified and unqualified, which are represented by "1" and "0", respectively. This is the ANN's output. For the input of an ANN, careful consideration is needed, and the information contained should be as complete as possible. Generally, it includes three parts: the parameters (x_1) continuously measured by the roller, the roller process parameters (vibration mass x_2, vibration frequency x_3, excitation force x_4, rolling speed x_5), filling parameters (type x_6, gradation x_7, particle shape x_8, moisture content x_9), a total of 9 inputs, as shown in Fig. 2.32. The key is to unify this input information into numerical data, especially by quantitatively describing the type, gradation, and particle shape of the filling material, drawing on the successful experiences from other industries.

Figure 2.32 illustrates the basic architecture of an ANN applied to compaction problems, which can be modified to achieve various applications. For example, changing the output of Fig. 2.32 to the test results of the plate load test (single output) becomes a prediction problem, but the input information is comprehensive and has a wide range of applications; if the output is modified to evaluate the compaction effect of the filling material (easy to compact, can be compacted, difficult to compact), it becomes a multi-classification problem.

In addition, using the learning architecture depicted in Fig. 2.32, the performance of the filler body (e.g., modulus) can be predicted based on the compaction process and filling parameters, thereby enabling a digital filler body that serves as the basis for CPS or a digital twin.

——*The similarities and differences between classification and clustering. Machine learning algorithms are mainly divided into supervised learning (regression and classification) and unsupervised learning (clustering), where supervised learning requires a large amount of known data for training, looking for the inherent regularity of data, and unsupervised learning does not require known data, automatically looking for the inherent regularity of data. Therefore, although both classification and clustering are forms of data partitioning, the former requires the support of known data, while the latter does not and is an automatic classification. Using classification or clustering algorithms for data preprocessing may be a good choice.*

The ANN model requires a large amount of known data to learn. The data comes from experimental roads and the accumulation of historical experience. Intelligent compaction provides convenient conditions for collecting compaction data.

(3) Knowledge can be stored in ANN—The combination of machine learning and expert systems

Expert systems simulate human reasoning ability. The primary issue is that they are unable to learn. The acquisition and conversion of knowledge rely on manual operations, which are time-consuming and laborious. Neural networks simulate the human brain and have the ability to learn. Combining ANN (or other intelligent technologies) with expert systems is a new development trend. In neural networks, rules can be learned, and knowledge is stored in the network's weights. A neural network can contain all the relevant knowledge of a problem, which is a manifestation of a set of rules, as shown in Fig. 2.33.

Figure 2.33 shows the technical approach of using an ANN to represent the compactability of fill material (the network structure needs to be carefully designed when implemented), which improves on Fig. 2.32. This network can form three rules, R_1, R_2, and R_3, corresponding to the three types of fill material: easy to compact, can be compacted, and difficult to compact. ANN outputs the compactability of the fill material based on the input information (each type can use 1 to represent "true" and 0 to represent "false"). The trained (learned) ANN model can obtain a set of stable weights, that represent knowledge (the same principle as expert scoring in statistical surveys, the weight determined by expert scoring represents expert's knowledge level), and form corresponding rules, which together form the ANN knowledge base. Additionally, if new training data becomes available, new weight distributions can be automatically generated from it, thereby updating the knowledge base.

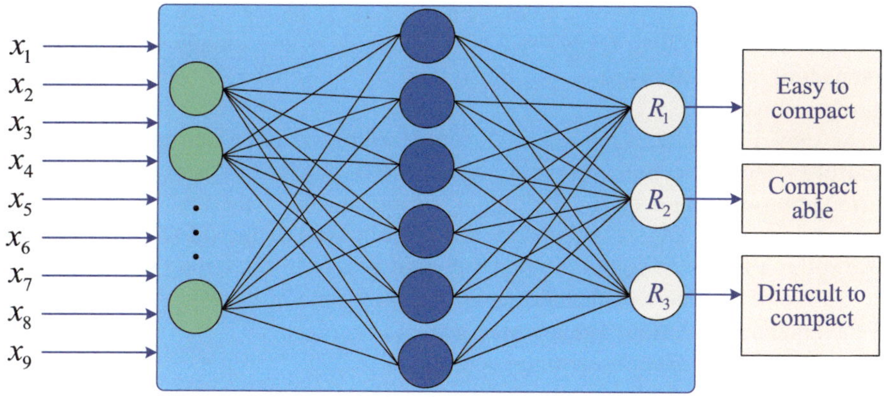

Fig. 2.33 Knowledge stored in ANN

Rules are explicit expressions, while ANNs are implicit expressions, and the two can also be used in combination. However, when using neural networks to express knowledge, the input knowledge must be converted into a digital form. As some knowledge may be difficult to digitize, neural networks are not omnipotent, and other machine learning algorithms can be considered. There are many intelligent algorithms currently, and the learning rules differ, which can be selected according to specific situations (this is also a promising research direction). In addition, knowledge with uncertainty can be represented by probabilistic methods, and fuzzy methods can handle knowledge with fuzziness. Combining emerging deep learning algorithms with expert systems is also worth considering.

Although expert systems lack universality and learning capabilities, they are highly useful in certain professional fields. When combined with machine learning, they possess multiple abilities, such as learning and reasoning. They are highly promising intelligent technology in the field of engineering construction, which deserves our attention. In addition, whether expert systems can be combined with the development of large models (such as ChatGPT and DeepSeek) is also worth studying.

(4) **Reinforced learning—Let the road roller work and learn at the same time**

Generally, road rollers operate according to a compaction plan, which is based on experience. After adopting intelligent technology, a large amount of training data is usually required. This data is derived from accumulated historical experience, encompassing many field trials, and the expertise of domain experts. Obtaining this data will require significant manpower and resources.

If a road roller can "learn on its own" during the compaction process, it can then carry out autonomous compaction operations and quality control. Reinforcement learning is an approach of this kind, where a compaction plan is not given in advance, and the road roller finds the best compaction plan (strategy) through continuous attempts during the compaction process. The basic principles of reinforcement learning are introduced in the introductory volume of the series; here, we primarily discuss how to apply them. The following is an example of a road roller compacting a specific type of fill, illustrating the basic idea of using this method during compaction.

As shown in Fig. 2.34, in reinforcement learning, a road roller equipped with a control system (perception terminal + software) is referred to as an agent, and the surrounding environment is called the environment. After the road roller issues a compaction action (A), it will change the current state of the fill (S) and give the road roller a feedback signal—called a reward or return (R), telling the road roller the degree of goodness of the current compaction state of the fill, as an evaluation of the compaction action. The learning goal is to achieve the maximum cumulative reward (similar to students earning good grades through learning).

Suppose the road roller's compaction operation (action) at the t moment (or the t_{th} time) is denoted as A_t, and the two feedback information—compaction state is denoted as S_t, and the reward is denoted as R_t. S_t and R_t will determine the

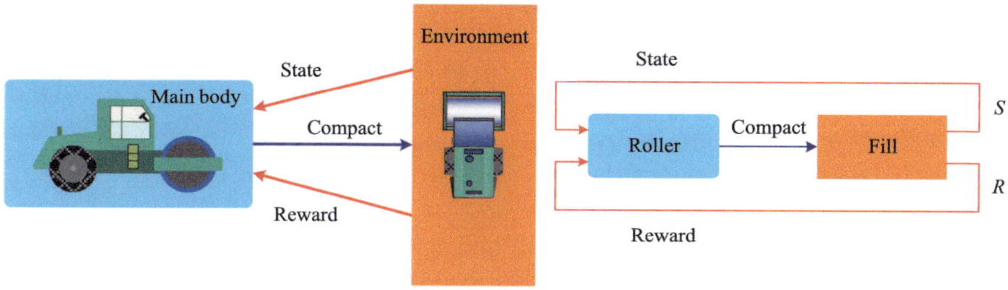

Fig. 2.34 Principle of reinforcement learning in compaction operations

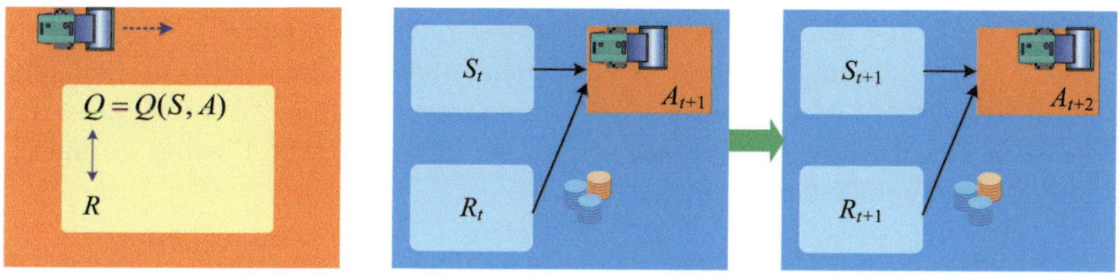

Fig. 2.35 The current state and reward determine the road roller's action at the next time

road roller's action at the next moment (or the next time) $(t + 1)$, A_{t+1} (continue compaction, change process compaction, stop compaction), action A_{t+1} will produce a new compaction state S_{t+1} and the corresponding reward R_{t+1}, they will determine the action at the next moment (next time), …, as shown in Fig. 2.35.

The purpose of reinforcement learning is to let the road roller learn on its own during the compaction process through continuous adjustment of actions and trial and error, to learn what kind of action to choose in what kind of situation, to maximize the accumulated reward (return), and to find the best compaction plan (strategy).

In reinforcement learning, action A, state S, and reward R are three key parameters that determine how to choose the best compaction plan. They are explained separately below.

The road roller may perform four types of actions (rolling pattern): continue compaction with the original process, increase process parameters (such as increasing vibration force) for compaction, decrease process parameters for compaction, and stop compaction operations, which can be represented by a_1, a_2, a_3, a_4, respectively, they constitute the set of road roller actions A.

There are many compaction states (S) of the fill, and appropriate parameters need to be selected for representation (see Sect. 4.2 later). For example, modulus E can be selected as the state variable, and different compaction states can be divided according to changes in E, such as E_1, E_2, E_3, …, forming a set of compaction states S.

The reward (return) R is a value set by humans, which is an evaluation of the action of the roller at time t (or the tth time), which can be represented by an integer or a decimal. This is similar to the principle of students taking test papers, where the score (or return) for each question is also set in advance by humans.

The task of reinforcement learning is to combine the rolling action, compaction state, and reward to enable the roller to find the optimal action strategy (rolling pattern) that maximizes the reward. This can be achieved using the Q-learning method (1989).

As shown in Fig. 2.36, combine state S and action A to construct a value function $Q = Q(S, A)$, which is used to evaluate the goodness of the roller in a certain state. The value function Q represents the expected value of the reward that can be obtained by taking action A under state S. According to $Q = Q(S, A)$ and Fig. 2.36, a Q value table can be constructed, as shown in Table 2.1. In this way, the Q value can guide the roller's roll, and the key is determining the Q value in the table.

In Table 2.1, the initial values of Q are randomly set, referred to as estimated values. The purpose of Q*-learning is to continuously update the Q values in this table to make it an accurate Q table that guides* actions. The following formula is usually used to update the Q value.

$$Q(S_t, A_t) \leftarrow Q(S_t, A_t) + \alpha[R(S_{t+1}) + \gamma \max_a Q(S_{t+1}, A_{t+1}) - Q(S_t, A_t)] \quad (2.10)$$

In the formula, $R(S_{t+1})$ is the reward value of the roller in state S_{t+1}; γ is the decay rate between 0 and 1; $\max_a Q(S_{t+1}, A_{t+1})$ is the maximum Q value selected in state S_{t+1}; α is the learning efficiency.

Additionally, a neural network can be utilized to generate Q-values. State S and action A are used as the input of the ANN, and the output is the Q value. Train the ANN parameters on known data to obtain a final Q *calculation model*, which can then be used for reinforcement learning.

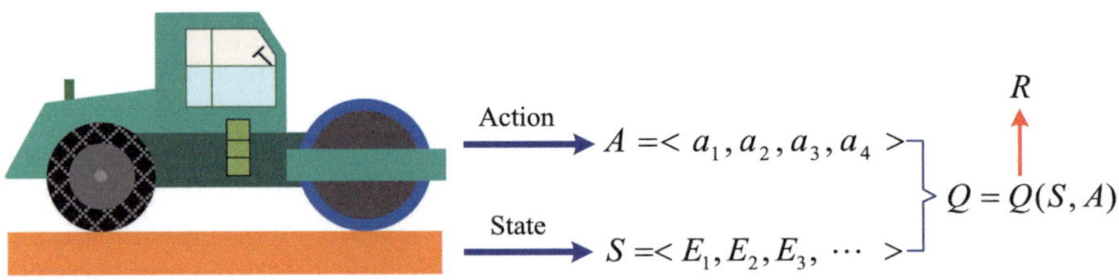

Fig. 2.36 Possible actions of the roller and the compaction state of the fill and their combination

Table 2.1 Q value

Q	a_1	a_2	a_3	a_4
S_1	$Q\ (E_1, a_1)$	$Q\ (E_1, a_2)$	$Q\ (E_3, a_3)$	$Q\ (E_1, a_4)$
S_2	$Q\ (E_2, a_1)$	$Q\ (2, a_2)$	$Q\ (E_2, a_3)$	$Q\ (E_2, a_4)$
S_3	$Q\ (E_3, a_1)$	$Q\ (E_3, a_2)$	$Q\ (E_3, a_3)$	$Q\ (E_3, a_4)$
...	...	...	...	...

The Q-learning algorithm is very simple: it involves performing different actions in each state, observing the returns (rewards) obtained, and iteratively updating the Q-value to reinforce the correct action. However, this method relies on the condition that the state set is known. If an unknown state occurs, this algorithm will not function and cannot predict such situations, so it lacks generalization ability and needs improvement, or alternative methods should be used.

2.7 The Core of Execution: Feedback Control

According to the technical characteristics of intelligent construction, which include "perception, analysis, decision-making, execution," the roller makes decisions based on feedback information. The last step is "execution"—continue rolling, stop rolling, or change the process of rolling, which is the "action" mentioned in the previous section. The purpose of execution is to roll, and its core is feedback control of compaction quality, which is also the fundamental task of intelligent compaction.

(1) **Starting from an example**

The concept of "control" originates in control theory, which involves applying specific effects to the controlled object to cause its behavior to develop toward the set goal. For example, the roller needs to roll along the set route. Among them, there is a simple control technology—by adjusting the steering drum. The actual driving trajectory is controlled to prevent it from deviating from the set direction, as shown in Fig. 2.37.

The example in Fig. 2.37 includes some common terms in control theory (see the control technology volume of the series). The operation trajectory (pre-set) of the roller (controlled object) is referred to as the input signal (target), and the actual rolling trajectory is the output signal. The two should be consistent. If they are not, a deviation will occur. The driver's eyes (observer) transmit the actual trajectory as feedback information to the input end, which compares it with the set trajectory. The driver (controller) then adjusts the steering drum (actuator) according to the size of the difference to eliminate the deviation.

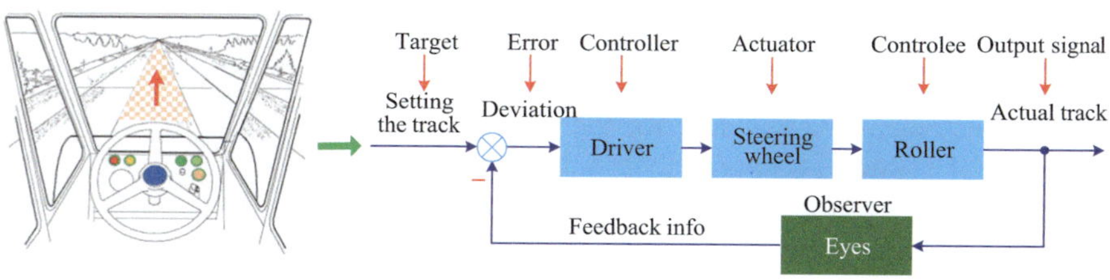

Fig. 2.37 Feedback control of the roller's rolling trajectory

The control problem of the roller's driving trajectory may seem simple, but it also encompasses a comprehensive set of control techniques, which serve as the foundation of autonomous driving technology.

——Control problems have existed since ancient times and later often appeared in the industrial field, thereby laying the foundation for automatic control technology. It was not until 1948 that Norbert Wiener published the book "Cybernetics," which shifted from general industrial control to encompass general control technology applicable to multiple fields, developing into a cross-cutting science with a methodological nature. Its basic concepts, basic ideas, and basic methods have penetrated various industries, spanning basic science, technical science, social science, biological science, and cognitive science, and emerging interdisciplinary subjects such as engineering control theory, social control theory, biological control theory, economic control theory, management control theory, etc. have emerged. It has now developed to the stage of intelligent control, which is specifically introduced in the control technology volume of the series.

Feedback is the "soul" of control technology, usually referring to the process of sending the output signal back to the input end. Feedback is categorized into two types: positive and negative. Positive feedback means that the feedback signal sent back to the input end is in the same direction as the input signal, which amplifies the deviation. Negative feedback, on the other hand, means that the feedback signal sent back to the input end is in the opposite direction to the input signal, thereby suppressing the deviation. Figure 2.37 shows negative feedback—comparing the actual output signal of the roller with the expected output signal (target), producing a deviation after the controller (driver's brain) processing, generating a control law, commanding both hands to manipulate the steering drum to eliminate the deviation. "Information, feedback, control" are the core elements of control technology, also known as the three pillars. They are interdependent and inseparable, and this is also true for intelligent compaction.

There are two modes for controling a system (object). One is open-loop control, that is, there is no reverse connection between the output and the input, and the output is controlled unidirectionally by the input; the other is closed-loop control, that is, there is a reverse connection (feedback) between the output and the input, and the output is controlled by the deviation between the input and the feedback signal. Obviously, the control mode of intelligent compaction is closed-loop feedback control.

(2) **System and Control**

Control involves the concept of the system (see the introduction volume of the series). The three key elements that describe the system's characteristics are input, output, and model. If a system is controlled, then the system itself is the controlled object, and the mechanism that implements control is called the controller (or control device), which primarily includes the observer, controller, and actuator. They together form the control system (see the control technology volume of the series), as shown in Fig. 2.38.

In Fig. 2.38, the observer is responsible for obtaining the output information of the controlled object and transmitting it to the controller; the controller is the brain

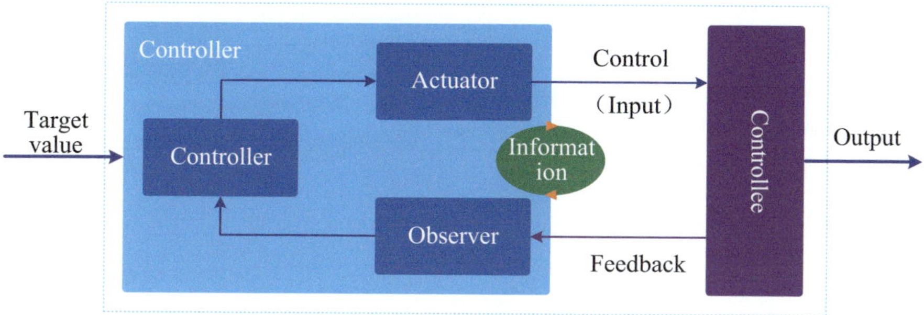

Fig. 2.38 Elements of the control system

of the controller, mainly responsible for receiving feedback information and giving target (reference input, which can be a given value or a given rule), and processing them into deviation (error) signals, then analyzing and deciding according to certain rules, generating control signals that change the input of the controlled object, and then transmitting them to the actuator; the actuator adjusts the input of the controlled object according to the control signal to keep the output near the target.

To control a system, you first need to understand its current state. You need to provide a mathematical model of the controlled system that allows you to control it effectively. Classical control technology employs ordinary differential equations to describe the system, primarily using single-input, single-output models; modern control technology, on the other hand, utilizes multiple inputs and multiple outputs and describes the system using state equations. These control technologies are very mature, and the challenge in engineering construction is primarily how to apply them; the key remains the accurate description of the system.

As the system (controlled object) becomes more complex, the model describing it becomes increasingly complex, posing new challenges for control technology. Intelligent control technology has evolved in this context, and technologies that utilize neural networks, expert systems, and genetic algorithms for control have emerged, which are still under development, as detailed in the control technology volume of the series.

It should be noted that although the control problems in different fields are different, they are all based on the same principle, which is also the basis for us to use the knowledge of other disciplines to deal with the problem of compaction quality control .

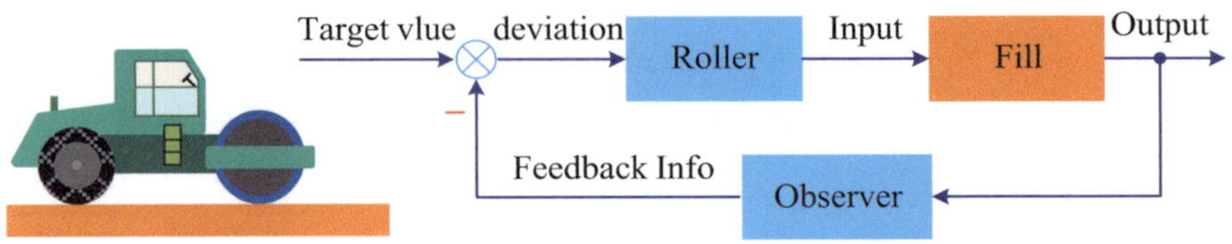

Fig. 2.39 Feedback control principle of intelligent compaction

(3) Control Principle and Control Method of Intelligent Compaction

In the context of intelligent compaction, "execution" refers to the roller's rolling operation (construction process), which ultimately affects in the feedback control of compaction quality. Figure 2.39 shows the control principle of intelligent compaction.

The fill here is the controlled object (the vibrating drum of the roller as an input device, which is also part of the controlled object), which partial differential equations can describe; the input (compaction) is the compaction force of the roller; the output is the performance parameters of the fill (such as modulus); the observer is the measuring device for intelligent compaction, responsible for continuously sensing compaction quality information; the controller (computer) located in the roller cab is the core of the control system, responsible for comparing feedback information with target values (compaction standards, such as $[E_R] = 80$ MPa), deciding the control plan based on the deviation, and issuing control instructions; the actuator is the adjustment mechanism for the roller's vibration process parameters, changing the compaction force according to the control instructions, thus allowing the roller to perform more effective compaction operations.

The feedback control process of intelligent compaction is not complicated. The roller applies a compaction force (input) to the fill. The filler's output is continuously measured by the measurement system, compared with the preset target value, and based on the deviation size; the vibration process parameters are adjusted (e.g., compaction force) to achieve consistency with the target value.

For intelligent compaction, from autonomously "perceiving" the output information of the filler to the "analysis and decision" of the information, these processes can all be completed automatically by the machine. Currently, there are three modes: manual control, automatic control, and intelligent control, as shown in Fig. 2.40.

Conventional rollers are manually operated to control the vibration process parameters (frequency, excitation force, amplitude, etc.). More advanced rollers can automatically operate process parameters, but there are still some issues with the adjustment range. Rollers that fully implement intelligent control are still in the early stages, but it is believed that they will develop quickly as the relevant foundations have been laid. The biggest problem currently is how to quantitatively adjust the compaction force.

Quality control is the foundation of engineering construction and the core of intelligent compaction. For intelligent compaction, the content of compaction

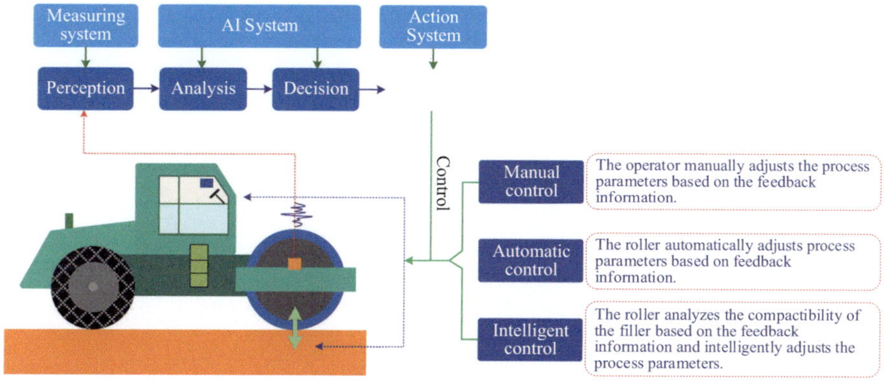

Fig. 2.40 Feedback control mode of intelligent compaction

quality control is very rich, including construction management information monitoring, compaction process, degree of compaction, compaction uniformity, control of compaction stability, and control of weak areas, etc., as detailed in the next chapter.

Chapter 3
Control Elements

Conventional compaction quality control relies on sampling detection, but the emergence of intelligent compaction has changed this situation. This technology, which integrates compaction and control, enables us to obtain more information, allowing for comprehensive quality control of the filler. Therefore, it is necessary to understand what control elements can be achieved through intelligent compaction and what the corresponding technical requirements are.

3.1 From Post-control to Real-Time Control

The birth of a product begins with design drawings, which involve first selecting suitable raw materials and then manufacturing them. The raw materials and the production process are the two key links that determine the product's quality. Similarly, for a "product" such as a fill structure, the raw materials consist of various fillers, and spreading and compaction are the production processes completed by construction machinery, as shown in Fig. 3.1. Therefore, filling and spreading compaction are the key links determining the quality of fill engineering and are also the focus of quality control.

Fill engineering is complex due to the source and type of fillers, making quality control of raw materials difficult. This can also be transmitted to the spreading and compaction stage, increasing the difficulty of quality control. For such engineering, the choice of control method is very important. Figure 3.2 compares point control and continuous control, highlighting the significant differences between the two.

Sampling detection is not only time-consuming and labor-intensive but also a form of post-control, making it difficult to handle problems in a timely manner. In many cases, the representativeness of sampling points is also problematic, as it falls under the "small data" control, and the control content generally encompasses only the "degree of compaction."

G. Xu et al., *Pioneer of Intelligent Construction - Intelligent Compaction*,
https://doi.org/10.1007/978-981-95-4184-3_3

Fig. 3.1 Formation pathway of fill structure

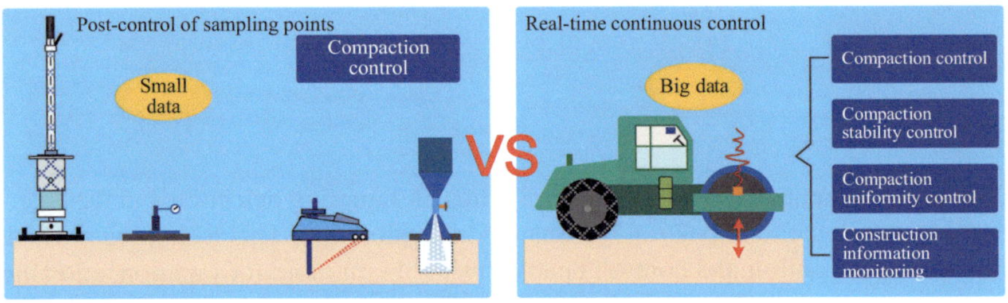

Fig. 3.2 Post-control versus real-time control

Intelligent compaction is continuously detected during compaction, enabling real-time, continuous process control. It can obtain information on all points on the compaction surface (including different compaction passes) for "big data" control. From this, many control contents can be derived (Fig. 3.2), which will be detailed later.

Fill engineering has evolved from point-post control to continuous real-time control over several decades and remains a frontier technology in this field, still in development.

3.2 Raw Material Control

The first step in controlling product quality is raw materials, which are the basis of process control. For intelligent compaction, if there is a problem with the filler material, even the best compaction technology will find it difficult to obtain a high-quality structure.

Due to the complexity of the raw materials used in fill construction, the key control points for each material differ. Therefore, it is essential first to understand the source of the materials and identify the key quality control points based on it.

(1) **Source of materials**

The raw materials for filler construction can be divided into two major categories, as shown in Fig. 3.3. The first category includes natural rock and soil materials, such

Fig. 3.3 Source of materials

as clay, gravel, and boulders, which are mainly used for various types of roadbeds (highways, railways, airports, and city roads). The second category includes artificially prepared materials, such as various sizes of stone, asphalt, and concrete, which are mainly used for the construction of base and surface layers (including airport runways).

For artificially prepared materials, from preparation to mixing, quality control is relatively strict. The most difficult thing to control is the roadbed fill material. These types of rock and soil materials not only have complex sources, but the quality standards set also have many loopholes, such as "requiring good gradation of coarse-grained fill materials," but there are no specific gradation requirements, nor are there regulations on how to prepare, making it impossible for construction units to operate specifically. In this case, various technical measures can only be adopted to ensure compaction quality. Below are a few control points for rock and soil materials.

(2) **Control points for rock and soil materials**

From a quality control perspective, factors that affect the compaction of rock and soil materials include moisture content, particle gradation, and particle shape, all of which are key points in quality control.

1) **Moisture content**

Materials that are sensitive to moisture content are mainly sticky, fine-grained materials (mainly clay). If the moisture content of these fill materials is not controlled near the optimum, it isn't easy to compact. Figure 3.4 illustrates the relationship between moisture content distribution and corresponding physical and mechanical properties, providing a basis for quality control.

According to Fig. 3.4, when the moisture content of the fill material deviates from the optimum moisture content, but the deviation is not large, the fill material will generally be compacted, but two situations will occur—when $w_1 < w < w_m$, the mechanical properties are high but unstable; when $w_m < w < w_2$, the mechanical properties are low and unstable. In addition, if the moisture content is too small, i.e., $w < < w_1$, the fill material will be difficult to compact, and at this time, water spraying treatment is needed; if the moisture content is too large, i.e., $w > > w_2$, the phenomenon of "spring soil" will occur, and at this time, it is necessary to carry out sun drying or add lime mixing to reduce the moisture content.

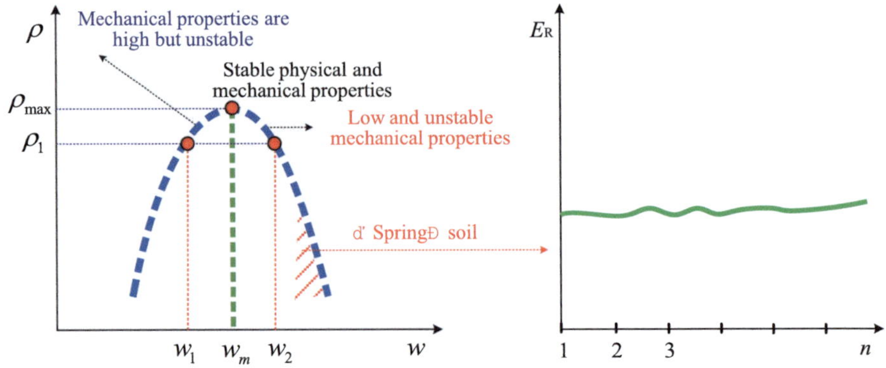

Fig. 3.4 Relationship between material moisture content and physical and mechanical properties

2) Gradation

For coarse-grained materials, whether they can be rolled into a structure with excellent performance mainly depends on their gradation and rolling process. Coarse-grained materials with poor gradation are difficult to roll into shape. Since there is currently no clear regulation on the gradation of roadbed fill materials, one can only explore on one's own. If you use the sieving exclusively, it will be very costly. The method of image recognition may be a shortcut, but it requires manual improvement of gradation.

Soil and stone mixed fill materials occupy a certain proportion in roadbed construction. According to practical experience, the ratio of soil to stone is about 3:7, and the structure formed by rolling has the strongest resistance to deformation. If it is mainly soil, the moisture content should be considered.

3) Particle shape

The particle shape of coarse-grained materials determines the relationship between particles and determines the quality of the filler body. In addition to good gradation, particle shape is also important, generally requiring clear edges and corners, which can increase the particle-to-particle interlock and form a more stable structure.

When using coarse-grained materials as roadbed fill, two main issues arise. One is whether it can be rolled into shape, which is related to particle shape and gradation; the other is whether the physical and mechanical properties of the formed structure are optimal, which is related to gradation (ratio of coarse to fine materials).

In Fig. 3.5, the particle shape of (*a*) is relatively smooth, and it is difficult to roll into a stable structure regardless of whether the gradation is good or not; the particle shape of (*b*) is sharp, but the particle size is single, and it is also not easy to form a stable structure; The particle shape and size distribution of (*c*) are relatively good, which can be crushed into a stable structure.

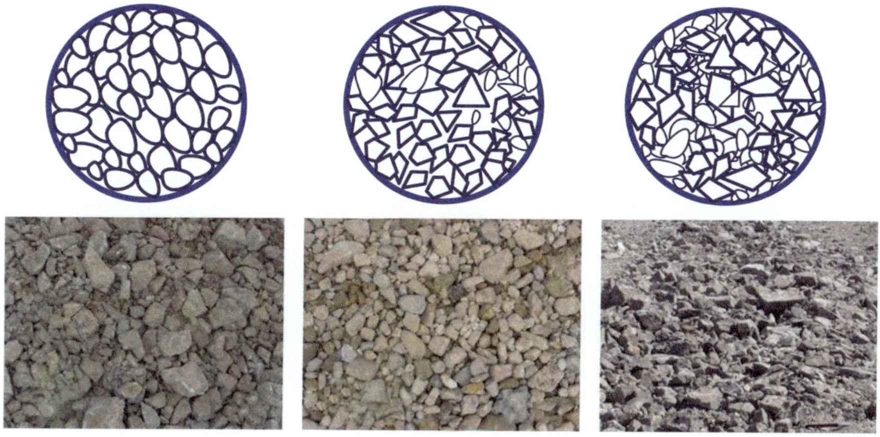

Fig. 3.5 The particle shape and gradation have a significant impact on the structure

3.3 Degree of Compaction Control

The degree of compaction refers to the extent to which the index representing the physical and mechanical state of the compacted layer reaches the target value during the filling and compaction process. If mechanical indicators (modulus) are used as control indicators, then:

$$E_i \geq [E] \tag{3.1}$$

where:

E_i is the continuous detection result of the i-th detection unit on the compaction surface, and the area of the detection unit is the product of the length of the vibrating drum of the roller L and the compaction distance (generally taken as 0.3–0.5 m);

$[E]$ is the target value, that is, the control standard.

In Eq. (3.1), the target value $[E]$ is determined by comparing the traditional point detection results, and it corresponds to the standard value of the traditional control index (such as K_{30}). See the later discussion for details. Figure 3.6 shows a schematic diagram of the compaction degree control method.

It should be noted that in intelligent compaction, a detection unit corresponds to a specific point that represents a comprehensive test value. However, this "point" has a certain area (0.5–1.0 m^2), just as in a plate load test, which is approximately 0.07 m^2. In addition, in the direction of the drum length L of the steel drum, the minimum resolution unit is the drum length, and it is impossible further to subdivide the response value of the steel drum. Theoretically, regardless of the number of sensors installed on the steel drum, their dynamic responses should be the same. Otherwise, it is not a rigid body.

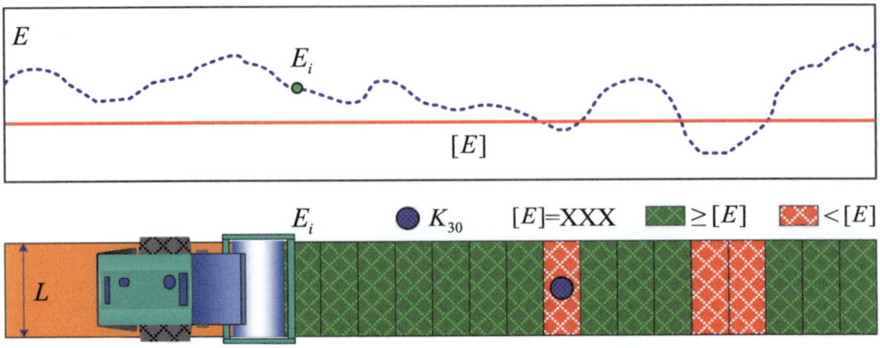

Fig. 3.6 Degree of compaction control

The concept of "degree of compaction (or compaction coefficient)" is a shortened form of "degree of compaction." Still, it is generally used specifically in the context of the dry density index system. Therefore, the degree of compaction mainly refers to the percentage of the field dry density that is equal to the standard dry density.

If all the detection results on the compaction surface meet the requirements of Eq. (3.1), it is undoubtedly ideal. Still, it is difficult to achieve in practice, and it is acceptable to have a small area below [E]. There is a pass rate issue.

The so-called pass rate refers to the percentage of detection units (compaction pass area) that meet Eq. (3.2) on the compaction surface, divided by the total number of detection units (total compaction surface area). It is generally required that the pass rate of the degree of compaction of the compaction surface reaches the specified requirements, that is:

$$\beta = \frac{\textit{Passed area}}{\textit{Total compacted area}} \times 100\% \geq [\beta] \tag{3.2}$$

where $[\beta]$ is the specified pass rate standard, which can be set according to the project level and technical requirements, and can generally be set to $[\beta] = 90\%$. The Chinese railway standard stipulates $[\beta] = 95\%$, which is higher than the regulations of other countries (generally 90%).

If the pass rate is set at 90% or higher, it means that there is still a 10% area that does not meet the requirements. 10% doesn't seem to be much, but don't forget that this is 10% of the entire compaction area. Taking a compaction surface of 100 m × 20 m = 2000 m^2 as an example, the non-compliant area is 200 m^2. If all or part of this area is concentrated, it will be very dangerous and will pose a threat to the upper structure and driving safety. Therefore, supplementary regulations are needed for the non-compliant area—they cannot be concentrated together, as shown in Fig. 3.7.

Currently, there are no specific requirements for assessing the concentration and dispersion of non-compliant areas, which involve issues of spatial variation and distribution. Generally, it can be judged based on the distribution map. Figure 3.7

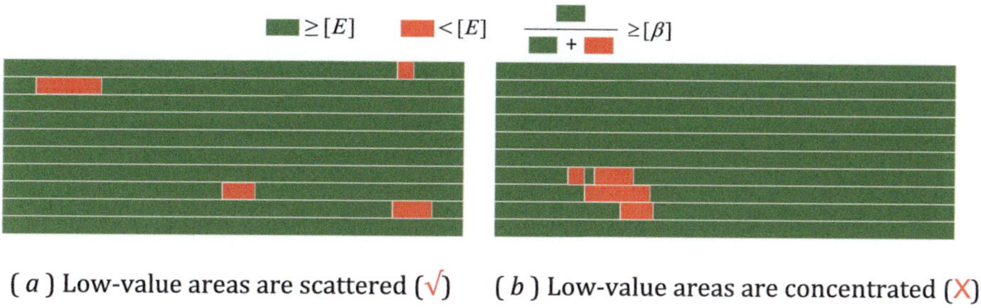

Fig. 3.7 Distribution requirements for low-value areas

shows two situations where (*b*) is unacceptable (but it should be related to the specific area size, generally requiring this low-value area to be 5–10 m^2), and necessary engineering treatment should be carried out.

In fill construction, the degree of compaction is the most important control element, as it bears the heavy responsibility of ensuring sufficient rigidity and strength to support the upper structure and vehicular loads. This is also the reason why traditional point control mainly detects compaction and modulus. However, controlling the degree of compaction alone is not enough. It is also necessary to control compaction stability and uniformity.

3.4 Compaction Stability Control

In many cases, even if the degree of compaction meets the specified requirements, the performance of the filler body will still change under load, indicating poor stability. Stability refers to the property that the physical and mechanical properties of the filler body do not change significantly over time.

During the compaction stage, compaction stability refers to the property in which the physical and mechanical properties (compaction state) of the filler body do not change with the number of compaction passes under certain compaction process parameters, which can be characterized in the following way.

$$\delta = \frac{\Delta E}{E_i^{(n)}} = \frac{[E_i^{(n+1)} - E_i^{(n)}]}{E_i^{(n)}} \leq [\delta] \tag{3.3}$$

where:

$E_i^{(n+1)}$—The continuous detection result of the *i* point in the $(n + 1)$th pass;

$E_i^{(n)}$—The continuous detection result of the *i* point in the *n*th pass;

$[\delta]$—Control accuracy, determined according to the project level, coarseness of fill material, and the condition of the roller, can be taken as 1–3%.

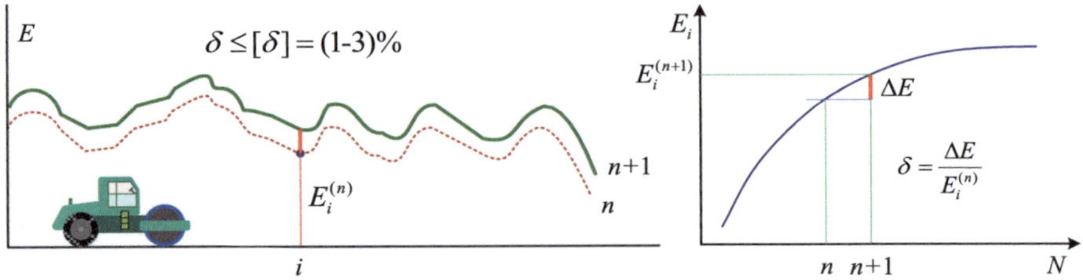

Fig. 3.8 Compaction Stability Control

$\triangle E$ is the key element to determine the stability of compaction. As the number of roller passes N increases, the state of the filler body transitions from loose to compact, and $\triangle E$ also changes accordingly. The changing trend can be used to judge the stability of compaction, as shown in Fig. 3.8.

Theoretically, when $\triangle E = 0$, it indicates that the compaction work of the roller under this process has been maximized, and the plastic deformation and compaction state (physical and mechanical properties) of the filler body no longer change. At this time, the compaction operation should be stopped. Whether it meets the specified requirements should be determined in conjunction with formula (3.1). In actual work, the requirement of $\triangle E = 0$ is too strict; the condition can be appropriately relaxed, and $\triangle E$ is controlled within the allowable range, which is why the control accuracy $[\delta]$ is stipulated.

For fill construction, stability is very important. Unstable structures will exhibit various problems under repeated vehicular loading, and fatigue damage is one of them. Therefore, during the construction period, the filler body should be compacted to a relatively stable state to the extent possible, and the judgment and control method is as outlined in formula (3.3). However, it should be noted that the compaction stability of different tonnage rollers or different compaction process parameters is not the same. For example, after the 16-ton roller is stable, the compaction state may still change under the action of the 18-ton roller.

Compaction stability has many uses. For example, with mixed fill materials or fill stones, due to particle size limitations, point detection cannot be carried out. It is common to use a certain tonnage (such as 18 tons) of vibrating roller to compact without drum marks at the end of compaction (i.e., proof rolling). Compacting without drum marks means no plastic deformation. However, for coarse-grained materials, if compacted to no plastic deformation (constant density), the internal particles may still adjust its packing; the resistance to deformation (modulus) will still increase, and compaction may not reach a stable state. Controlling compaction stability is crucial to eliminate this impact.

3.5 Compaction Uniformity Control

Highways and railways are a type of "belt" structure that stretches for hundreds of kilometers and comprises multiple structural layers. Although the physical and mechanical properties of different structural layers are different, the physical and mechanical properties of the same layer should be the same. As shown in Fig. 3.9, in the z direction, the modulus decreases gradually from the top down, but on the x–y plane, the modulus at each point should be the same (when z is constant). Otherwise, various problems may occur, such as localized early damage and uneven vehicle support.

Uniformity is a unique requirement for road and railway engineering structures, as it determines whether the structure can evenly support the upper structure and ensure safe driving. So, how can we achieve and ensure this uniformity? This heavy burden falls on the construction phase (product manufacturing phase), which must be addressed from both the raw material and the construction process. The raw material part encompasses material design, preparation, mixing, transportation, and other related processes. Here, we primarily discuss control issues in the construction process.

The fill structure is built layer by layer. Each layer undergoes two steps: spreading and rolling, with rolling as the final step and a crucial one for controlling the compaction uniformity (although the spreading step is also very important). Controlling uniformity during the rolling process means ensuring the distribution of physical and mechanical properties across each layer is uniform; however, there is currently no unified control standard.

To what extent should the uniformity of compaction be controlled to be considered qualified? Controlling uniformity requires a principle. That is, the variation of the filler body performance cannot have too much impact on the upper structure and driving. In analyzing and evaluating this impact, there is currently no unified regulatory framework. Highways and railways, as well as flexible pavements and rigid pavements, ballasted tracks, and ballastless tracks, all have different requirements for uniformity (also known as variability), which still need to be studied. Empirical methods are still primarily used for ensuring uniform compaction. Here are two simple judgment methods.

(1) **Mathematical statistical method**

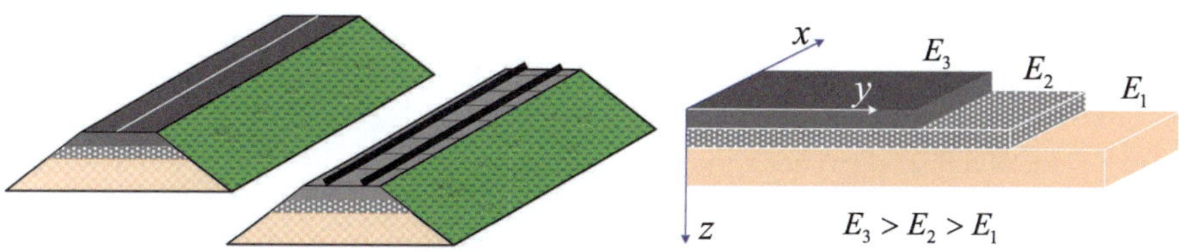

Fig. 3.9 The meaning of uniformity of highway and railway structure performance

The mathematical-statistical method is widely used to control quality in all walks of life. Based on the continuous, real-time detection data, relevant indicators are calculated using the mathematical-statistical method, enabling not only the assessment of uniformity but also the identification of problematic data and their distribution.

1) **Use the coefficient of variation to evaluate uniformity**

The coefficient of variation (C_V) is an important indicator of data variability and can be used to assess compaction data variability and to understand the overall variability of filler body performance. The coefficient of variation can be calculated as follows.

$$C_v = \frac{s}{\overline{E}} \tag{3.4}$$

where s is the standard deviation of the sample data, $\overline{E}$ is the mean of the sample data (here, the modulus E is used).

Regarding the relationship between the coefficient of variation and uniformity, there is no corresponding technical standard; therefore, a reasonable value can be determined through investigation and testing.

Although the coefficient of variation can evaluate the size of variability in compaction quality, it cannot determine the location of the variability or recommend engineering treatment. Therefore, it is also necessary to determine where the variability occurs.

2) **Distribution range of uniformity**

There is a concept in mathematical statistics known as a confidence interval, which is used to identify data that does not meet the requirements. We can use this concept to classify compaction data (uniform and non-uniform). For the continuous detection data E_i at a given point, if E_i meets the following requirements, it belongs to the uniformity category.

$$E_i \in [\overline{E} - Zs, \overline{E} + Zs] \tag{3.5}$$

where Z is the fluctuation coefficient, determined according to a certain assurance rate, in actual work, specific regulations need to be established for the fluctuation coefficient based on the type of fill, the level of project construction, and other relevant factors. For example, for highways, you can take $Z = 1.645$, the corresponding assurance rate is 90%.

Equation (3.5) is also known as the "3σ rule." In many experimental works, this method is used to find the outliers in the data. If E_i is not within the range shown in Eq. (3.5), whether it exceeds the lower limit or the upper limit, it is considered a non-uniform point, as shown in Fig. 3.10.

In practice, it is found that using the above method to control uniformity, results in most fill detection being non-uniform. For the non-uniform part, engineering treatment measures can be implemented in the low-value area; so what about the

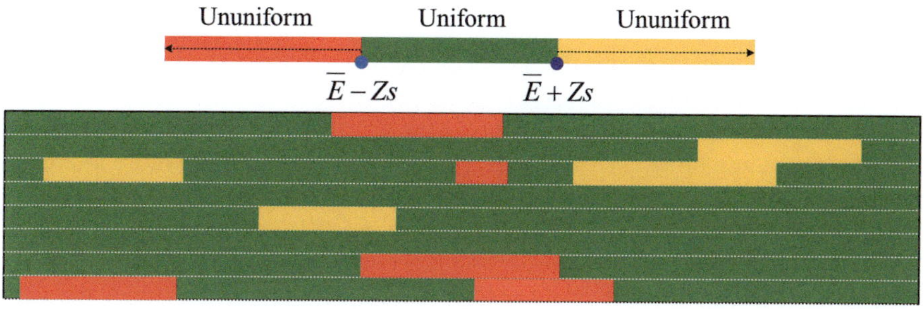

Fig. 3.10 Distribution range of compaction uniformity

high-value area? It is currently difficult to answer this question. Therefore, the current control principle is only to control the low-value area.

(2) **Recommended low-value control method**

If only the points below ($\overline{E} - Zs$) in Eq. (3.5) are considered non-uniform. It is low-value control. To simplify the operation, a simpler method is recommended below, which is also the method stipulated in the Chinese railway and highway industry standards, as shown in the following formula.

$$E_i \geq \eta \cdot \overline{E} \tag{3.6}$$

where:

η The correction factor is determined by the type of fill and the level of the project, and it is the level of variability control. The Chinese railway standard stipulates = 0.80, which can be referred to as shown in Fig. 3.11.

The purpose of controlling, according to formula (3.6), is to ensure that there are no areas with a compaction status that is too low to prevent uneven settlement. Engineering practice has proven that this control method is achievable at the current

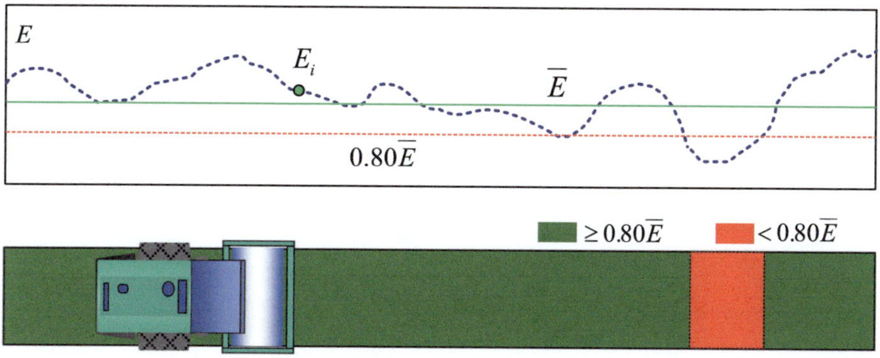

Fig. 3.11 Uniformity control of compaction

level of construction. If there is no significant variation in the fill, and the compaction is carried out according to the specified compaction process, it can generally meet the requirements.

The uniformity of compaction depends on the distribution of the fill and the control of compaction, among which the characteristics of the fill itself are more important and are the main reasons for variability, including variation in the moisture content of sticky fine-grained materials and gradation variation in coarse-grained materials. Controlling uniformity during compaction can only address problems arising from compaction and cannot completely resolve unevenness issues caused by the variability of the fill itself; however, it can identify these problems and provide a basis and specific location for engineering treatment.

3.6 Construction Information Monitoring

With the popularization of informatization and networking, construction information monitoring technology has developed well. For example, real-time tracking of construction progress and material input can be displayed in real-time on the large screen of the information management platform. There is also a significant amount of construction information involved in the compaction process. Although it is auxiliary information for intelligent compaction, it is also very important and can help improve project management.

Figure 3.12 illustrates commonly used construction information related to compaction operations. Most of this information can be obtained using satellite positioning technology. With this information, effective compaction management can be achieved. These pieces of information are closely related to project management, and while they seem simple, they are not. There is an inherent logical relationship between the information, and mastering this information can enable targeted project management.

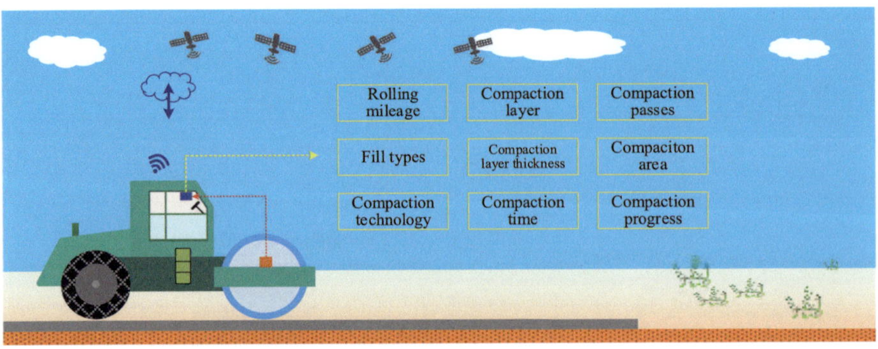

Fig. 3.12 Content of construction information monitoring

Among these pieces of information, the type of fill and compaction process is very important and necessary data for machine learning and expert systems in intelligent compaction, and attention should be paid to the accumulation.

3.7 Complementary Advantages: Coordination of Continuous Control and Point Control

Although intelligent compaction can continuously measure the modulus of the filler body, it does not mean completely abandoning point sampling detection. The two can complement each other's advantages, coordinate control, and convert random sampling into fixed-point detection, thereby reducing the number of point detections. For this, it is necessary first to divide the compaction surface into compaction status categories, determine the weak compaction area and specify the position of the fixed-point detection.

(1) Division of compaction status of the compaction surface

The compaction status of the compaction surface is the collection of compaction data (such as modulus) at each point. The left figure in Fig. 3.13 shows it in digital form (not all shown), but it is not very intuitive and requires a change in the display method.

Visualizing the compaction status of the compaction surface is a good choice. The compaction data can be graded at certain intervals (such as 10 MPa), and the same level is categorized and displayed in the same color. This is the division of the compaction status, as shown in the right figure of Fig. 3.13.

It should be noted that the compaction status distribution map and the compaction degree distribution map are two different diagrams. The compaction degree distribution map categorizes all compaction data into two categories based on the target value [*E*]. It represents them in different colors (see Fig. 3.7). The compaction status distribution map, on the other hand, categorizes all compaction data into multiple intervals and represents them in different colors. The compaction status distribution map serves as the basis for identifying areas with weak compaction.

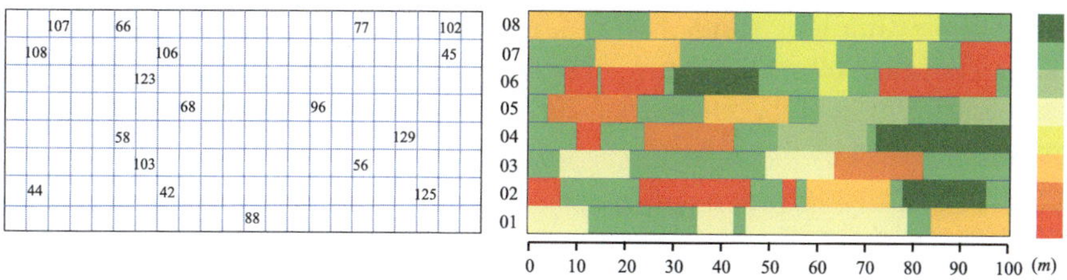

Fig. 3.13 Two representations of the compaction status of the compaction surface

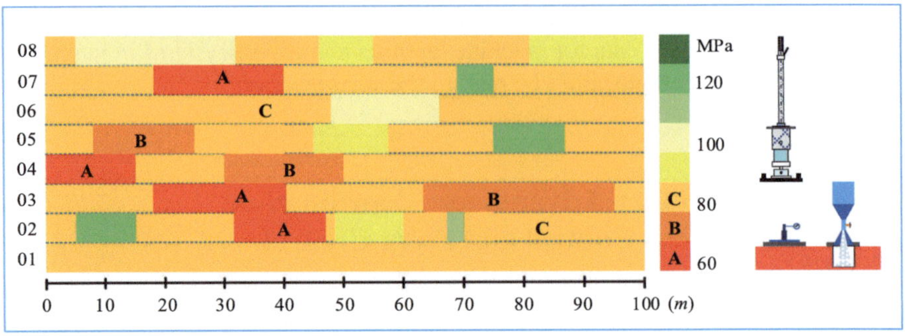

Fig. 3.14 Fixed-point step-by-step detection control

(2) Weak compaction area and fixed-point detection

The weak compaction area is the area with relatively low compaction, a relative concept. Suppose there is a certain difference in the compaction data on the compaction surface. In that case, there will be areas with relatively small data, but not necessarily less than the target value $[E]$.

For all compaction data on a compaction surface, after appropriate grouping, the low-value area indicates a relatively weak compaction quality. Starting from the lowest value group, it can be successively referred to as the weakest area A in terms of compaction quality, the next weakest area B, and so on, as shown in Fig. 3.14. The position of point detection is chosen in these areas, which represents an improvement over sampling-based detection and can also be referred to as the weak area control method.

Due to the randomness of the sampling point selection, the inspection point may not be selected in the worst-compaction state (even if it is, the distribution range cannot be determined), which may cause a "missed inspection" phenomenon. This increases the potential for non-compliance with compaction quality standards, whereas fixed-point control avoids this risk.

During fixed-point detection, the test should first be carried out in the area with the lowest compaction status (Area A in Fig. 3.14), and the connected area should be at least 5.0 m^2. If it is qualified, the entire compaction surface is qualified. Otherwise, point detection should be carried out in the next weakest area (Area B) with a connected area not less than 5.0 m^2, …, until a qualified area is detected. In this way, all areas suitable for engineering treatment can be identified. The combination of continuous control and point control can each play to their strengths and jointly complete the task of quality control.

——*The modulus obtained using the vibrating roller, compared with the plate load test, is relatively "rough" in terms of calculation models, loading equipment, loading methods, and measurement. The test results obtained by different rollers will not be the same. Under the current circumstances, it is still necessary to use the plate load test results for "calibration," which is also an important reason why point detection cannot be eliminated (though the number of point detections can be reduced). Suppose the vibration parameters of the roller can be standardized, and its calculation model, loading method, and measurement technology are unified. In that case, the detection accuracy can be improved, and the detection results will also be unified. In the future, canceling point detection may be considered a direction for development.*

Chapter 4
How to Implement

The relevant theories and control elements of intelligent compaction have been introduced earlier. This is a very practical technology that focuses on application and solving practical problems. This chapter will introduce how to apply intelligent compaction technology in engineering construction, based on the technical characteristics of "perception, analysis, decision-making, execution," demonstrate the working process of intelligent compaction, and finally provide a simple application example.

4.1 Understanding the Control System

To achieve intelligent compaction, the roller, as the primary component, should be equipped with the necessary hardware and software. This is the basic requirement for items to achieve intelligence (see the introduction of the series). According to the method of realizing intelligence in items, computers, sensors, and software (including control software and algorithm software) need to be integrated into the roller (Sect. 1.7). They are the key to forming the control system and the core of giving the roller the ability of "perception, analysis, decision-making, execution," as shown in Fig. 4.1.

(1) **Composition and Requirements**

The control system comprises four basic components: the controlled object, the observer, the controller, and the actuator (Fig. 2.28). For intelligent compaction, the control system is primarily composed of a fill, a roller, sensors, and a host (computer and software), as shown in Fig. 4.2.

In the intelligent compaction control system, the controlled objects include the fill and the roller (mainly the vibrating drum). There is interaction between the two, and the strength of their interaction reflects the change in the mechanical properties of the

G. Xu et al., *Pioneer of Intelligent Construction - Intelligent Compaction*,
https://doi.org/10.1007/978-981-95-4184-3_4

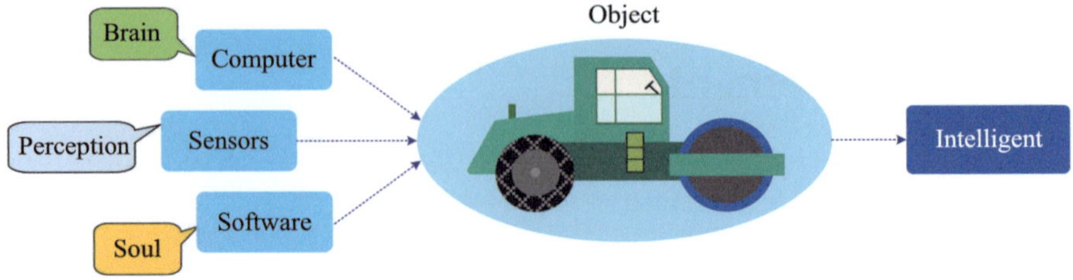

Fig. 4.1 The way to realize intelligence in rollers

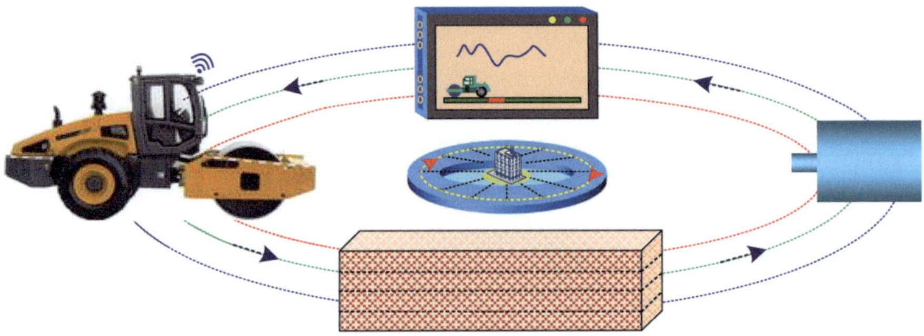

Fig. 4.2 Basic composition of intelligent compaction control system

fill (under the condition of a certain excitation force). The response of the vibrating drum can reflect this change; the observer is the sensor, which is responsible for sensing the response signal of the vibrating drum and transmitting it to the host; the host contains data collection devices, controllers, control software, and algorithm software, etc., responsible for receiving the signal from the observer and processing and storing it, analyzing the compaction information, making compaction process decisions, and issuing instructions on how to execute (control instructions); the actuator is the vibration parameter adjustment mechanism of the roller, responsible for adjusting the compaction action (by changing the input to the fill).

In the intelligent compaction control system, the software is the "soul." From the observer's perception of information to the controller's control instructions, all operations are controlled by the software. If there is an information platform, you can also collect scattered data, establish data-sharing mechanism, and conduct comprehensive data utilization on this basis, which also serves as the foundation of the Internet of Things in engineering.

The control system is composed of multiple parts (subsystems), each of which has specific requirements. Among these, the most important requirements are those for the roller and the measurement system (sensor + data collection device, also known as the perception terminal), which involve ensuring the accuracy of the compaction data.

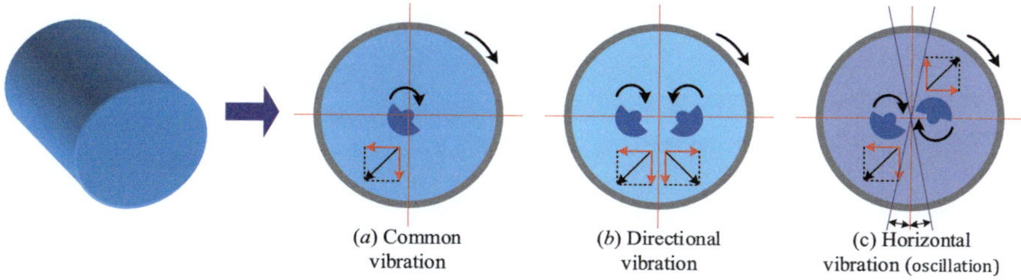

Fig. 4.3 Modes of action of the roller drum

1) **Requirements for the roller**

In loading, the roller's action mode has a great impact on the continuous measurement results (the vibrating drum's response). There are currently three modes of action, as shown in Fig. 4.3. If the control index utilizes mechanical quantities (such as modulus), because the calculation model employs a vertical acting force (Sect. 2.3), then the directional vibration mode is the most suitable. Ordinary vibration introduces a certain horizontal force that affects the test results. The horizontal vibration (oscillation) mode of action is not applicable to the current mechanical model and should be modelled separately.

In addition to the mode of action, the fluctuation of the excitation force of the roller (essentially, the fluctuation of the frequency) will also affect the test results. This problem is common in early vibrating rollers (Fig. 2.4). Therefore, the roller's vibration frequency must be controlled within the allowable fluctuation range. Generally, it should not be greater than (0.5–0.6) Hz. This requirement is similar to the loading requirement for a plate load test, which is determined by the fill's inherent characteristics.

As the executing agency, most current vibratory rollers employ manual methods (driven by the driver) to adjust the vibration process parameters. Some rollers with directional vibration can achieve automatic control (unmanned frequency and amplitude adjustment, but the adjustment range still needs further research). Intelligent control will be one of the main development directions of construction machinery in the future.

2) **Requirements for the measurement system**

In the intelligent compaction control system, sensors and data acquisition devices (including software) belong to the measurement system (modernly referred to as the perception terminal), which is also one of the main parts of the control system. There are currently various measurement systems with different functions. Still, there is a basic requirement: it must correctly perceive the compaction quality information of the fill.

How do we determine whether the measurement system can accurately assess the compaction quality? Generally, comparative tests are used for verification, which

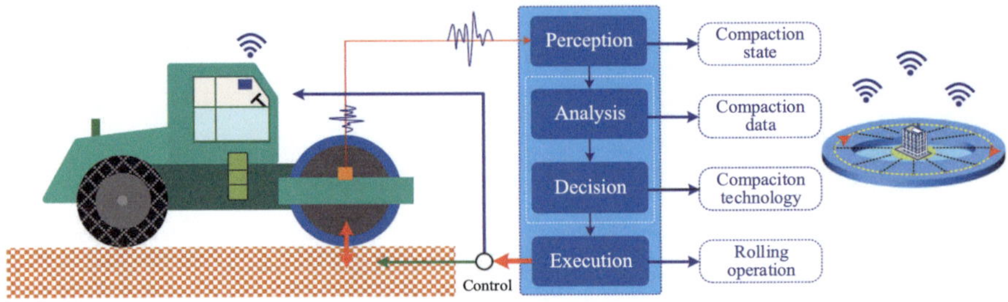

Fig. 4.4 Working process of intelligent compaction control system

assesses the correlation between continuous and point detection results. The correlation coefficient must be greater than 0.70 to be considered valid. For specific operations, see the engineering examples later.

(2) **Working process**

The working process of the intelligent compaction control system is illustrated in Fig. 4.4, which clearly outlines the key steps of "perception, analysis, decision-making, execution."

It should be noted that, if it is a more advanced control system, the above steps are all automatically completed without minimal human intervention. At present, most control systems can automatically perform the "perception, analysis, decision-making" steps. In contrast, the "execution" part (mainly feedback control) has not yet been fully automated with intelligence.

4.2 Perception of Compaction State

"Perception" of the compaction state of the fill is the first step of intelligent compaction, mainly completed by the measurement system (Sect. 2.5). With the development of electronic measurement technology, there are not many problems with the automatic measurement and processing of signals. The key issue is the clear understanding of the fill's compaction state.

(1) **Compaction state**

The compaction state refers to the physical and mechanical characteristics of the fill during the rolling process, encompassing a set of physical and mechanical parameters that reflect compaction quality. During the rolling process, density ρ and modulus E are usually taken as state variables, and they constitute the compaction state set $X(i)$ of the fill.

$$X(i) = \{x_1(i), x_2(i)\} = \{E(i), \rho(i)\} \tag{4.1}$$

In the formula, i represents the number of rolling passes (or time). ρ and E are variables during the rolling process, and they change nonlinearly with changes in compaction force. The current intelligent compaction technology (levels 3–4) primarily detects changes in the mechanical properties of the fill (modulus), while changes in density must still be detected using traditional methods.

(2) **Signal processing and control indicators**

The sensor installed on the vibrating drum of the roller (the current MEMS sensor has an embedded microcomputer and can output digital signals, becoming a perception terminal; see the Perception and Data Volume of the series) can perceive the vibration signal of the vibrating drum in real-time (displacement, speed, acceleration). The vibration signal carries information on the compaction state of the fill and the vibration parameters of the roller (Fig. 2.33), resulting from the interaction of the compaction force and the fill's resistance. The method of obtaining control indicators that reflect the compaction state through technical processing of the vibration signal is very critical. Different signal processing methods yield different control indicators.

The early method used spectral analysis of the vibration signal to represent the change in the compaction state with the harmonic ratio (f_2/f_1); intelligent compaction technology above level 3 indirectly obtains the displacement w and compaction force Q of the fill through the analysis of the vibration signal, thereby identifying the mechanical indicators of the fill (such as modulus, see Chap. 2), which can directly reflect the change in the compaction state, as shown in Fig. 4.5.

Using the modulus as an indicator of the compaction state is undoubtedly the best choice; however, there are also some technical issues. For example, the displacement w and compaction force Q of the fill are unknown and cannot be directly measured, requiring considerable technical assistance. In addition, when the roller bounces, the modulus calculation also encounters major problems.

(3) **Coupling and bouncing**

When calculating the modulus, it is assumed that there is a close contact (coupling effect) between the vibrating roller and the fill so that the displacement of the

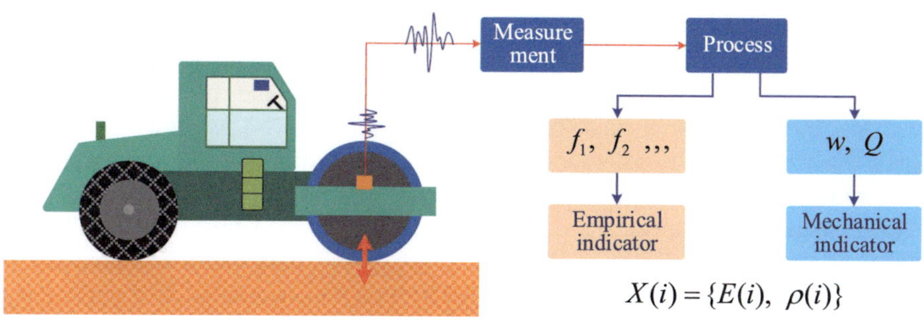

Fig. 4.5 Different processing methods of vibration signals yield different control indicators

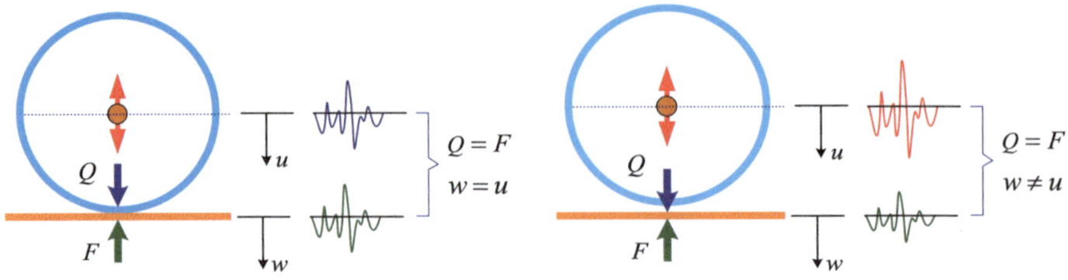

(*a*) Coupling (*b*) Collision

Fig. 4.6 Coupling and bouncing of vibrating drum and compaction surface

vibrating drum and the deformation of the rolling surface are coordinated. If bouncing (collision effect) occurs, the two are not coordinated, as shown in Fig. 4.6.

When the vibrating drum bounces, its displacement is relatively large, whereas the compaction surface's displacement is relatively small (due to its high stiffness). The earlier mechanical coupling model will no longer be applicable, and it is feasible to switch to a collision or impact model. At this time, the mechanical indicators obtained can be referred to as the "impact modulus," which also reflects changes in the compaction state, as detailed in reference [3].

— —*Perception technology is an integration of multiple disciplines. Although the electronic measurement system automatically completes the "perception" of intelligent compaction, it is very simple for users. However, for researchers, it is necessary to master knowledge in various fields to grasp it. The foundation of perception is not simple; the hardware foundation involves electronic information technology, and the foundation of perception methods involves knowledge such as dynamics. There are two other volumes in the series that specifically introduce this knowledge.*

Correctly perceiving the compaction state of the fill is just the first step in intelligent compaction work. The next work will be more challenging, focusing on analyzing compaction data.

4.3 Analyzing Compaction Data

Compaction data refers to the form of the compaction state data, which reflects the performance of the fill obtained by the roller. This data is the result of continuous detection, including data from different compaction passes (N), as shown in Fig. 4.7. This mainly refers to the collection of modulus under different compaction passes, and the acquisition of density data requires other methods.

Compared to spot-check data, compaction data is considered "full data," also known as "engineering big data." It is because of this data that a comprehensive analysis of the compaction degree, compaction stability, and compaction uniformity of the fill can be carried out, as shown in Fig. 4.8. This is something that spot-checking can hardly achieve.

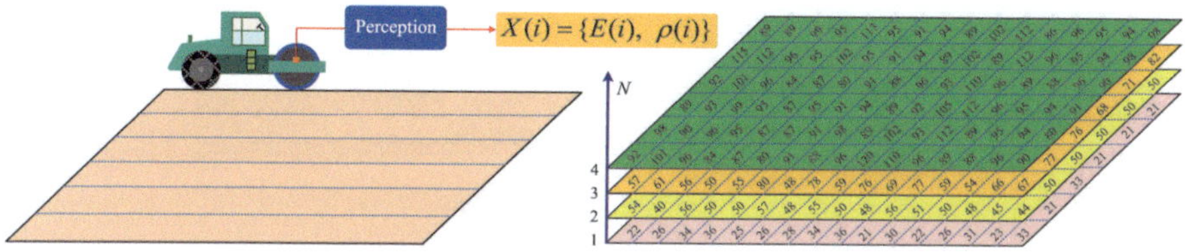

Fig. 4.7 Data form of compaction state

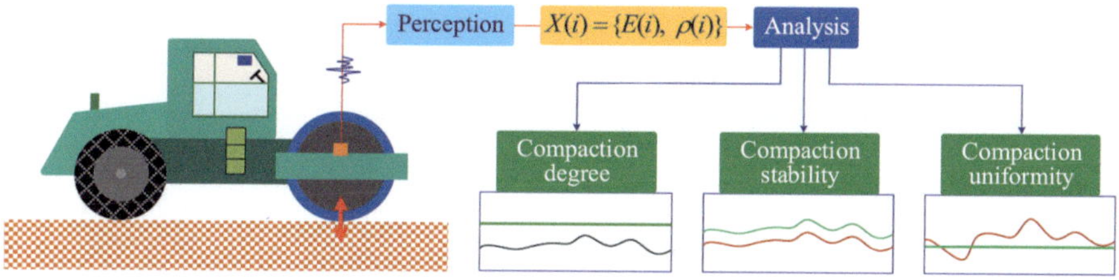

Fig. 4.8 Analysis content of intelligent compaction

(1) **Analysis of roller track**

At the start of the compaction operation, the intelligent compaction control system also begins to work. Before this, it is necessary to determine the target value of the compaction degree $[E]$, the pass rate $[\beta]$, the control accuracy of the compaction stability $[\delta]$, and the control level of the compaction uniformity η. See the relevant provisions in Chap. 3.

When the observer perceives the compaction data in real-time, the first thing the controller needs to analyze is the compaction degree, and it is judged whether it meets the specified standard $[E]$ according to formula (3.1); after a track is compacted, the average value of the track data can be obtained for compaction uniformity analysis; from the second pass, compaction stability analysis can be carried out. Figure 4.9 gives the specific analysis content and order.

The control system installed on the roller can automatically perform the above analysis during the compaction process and display the results on the screen for real-time decision-making and control.

Figure 4.9 gives an analysis of the roller track. If there is enough compaction data (at least one compaction operation is completed on the compaction surface), the compaction surface can also be analyzed, as detailed below.

(2) **Analysis of compaction surface**

From the second pass of the compaction operation, the entire compaction surface can be analyzed to assess the overall compaction quality. In addition to compaction degree, compaction stability, and compaction uniformity, the analysis of the compaction surface should also provide the distribution of the compaction state,

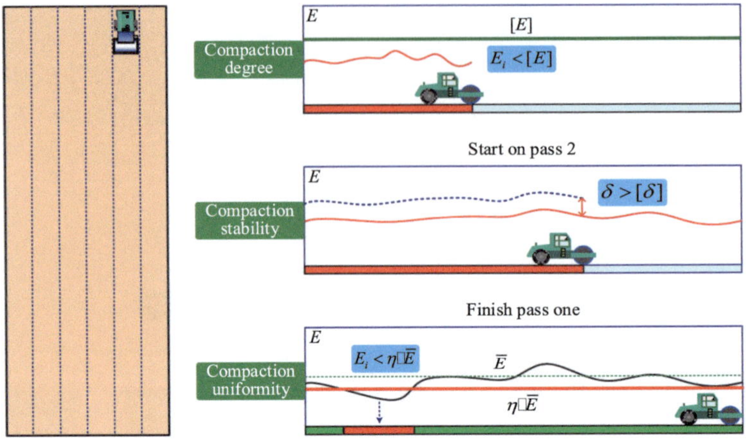

Fig. 4.9 Analysis of roller track

which serves as the basis for identifying weak compaction areas. How to operate: see the methods given in Chap. 3.

Figure 4.10 shows a schematic diagram of the compaction surface analysis (performed during a specific compaction pass), which is automatically completed by the machine. The compaction degree is divided into the target value [*E*]. Red represents not up to standard, and its pass rate does not meet the specified requirements; the compaction stability is analyzed by the track as a unit, red represents that the compaction data difference between the front and back of this place is large, which is unstable; the compaction uniformity of the compaction surface is somewhat different from the uniformity of a single track, mainly because the average value of the compaction data is different, the former is the average value of all data on the compaction surface, and the latter is the average value of the data on the track, red represents less than 0.8 times the average value. The analysis of these items is basically the same as the analysis of the track; the amount of data is just different.

The distribution of the compaction state on the compaction surface is very important. It is the grouping of all compaction data on the compaction surface. Different colors represent different compaction states (red represents the smallest value; see Sect. 3.7), which is an important basis for selecting spot-check points.

It should be noted that the three control elements of degree of compaction, compaction stability, and compaction uniformity are synchronously controlled during the rolling process, with no distinction in importance. However, the focus may vary depending on the type of fill materials. For example, the current focus of the stone fill subgrade is on analyzing its compaction stability as a replacement for the drum track method.

If the fill material is good, it is relatively easy to analyze the compaction data. Automatic analysis can be achieved using ordinary methods; there is no need to resort to intelligent methods. However, suppose the compactability of the fill material is poor (possibly due to gradation, particle shape, moisture content, etc.). In that case,

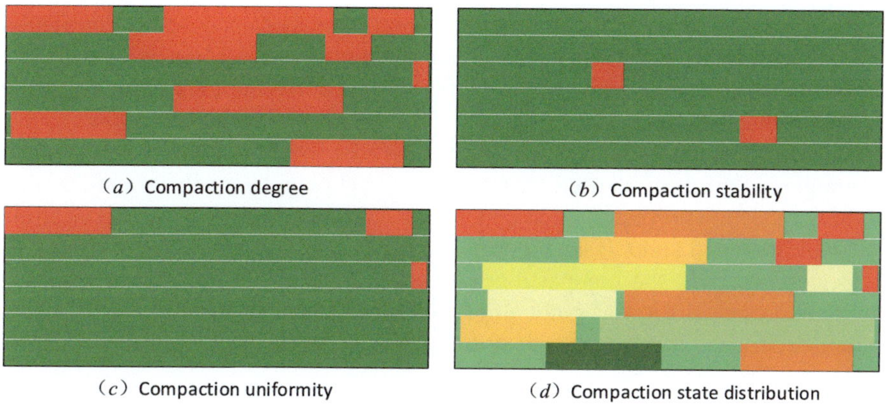

Fig. 4.10 Analysis of the compaction surface under a certain compaction pass

there will be many problems with the compaction data, and multiple analysis methods and treatment plans need to be adopted. Some discussions are conducted below in conjunction with the decision-making process.

4.4 Decision on Rolling Plan

The analysis is the cause. The decision is the result. The two complement each other and are indispensable. During the rolling process, the machine (specifically, the computer in the control system) makes decisions based on the analysis results to determine the rolling plan (or rolling strategy), as shown in Fig. 4.11.

The so-called compaction process refers to the parameters that affect compaction quality, such as the roller's self-weight, vibration mass, excitation force, vibration frequency, amplitude, and travel speed. A good compaction process should match the fill material and compaction state to achieve good rolling results.

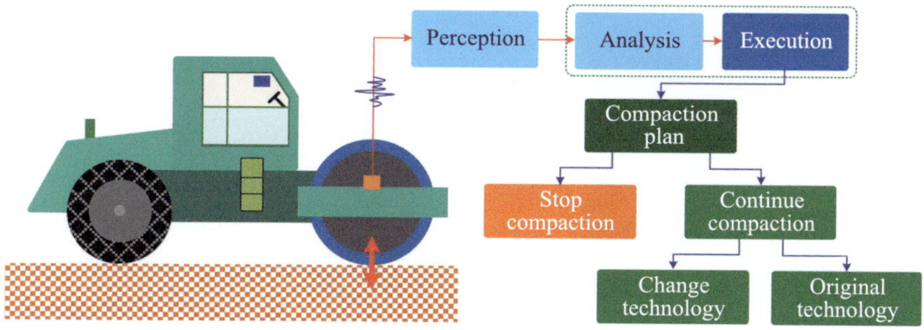

Fig. 4.11 Decision content of intelligent compaction

If the compactability of the fill material is good, it is easy to analyze the compaction data and determine the rolling plan. If the method given in Chap. 3 is followed, automation can be achieved without too much intelligent technology. However, when the compactability of the fill material is poor, the situation is more complicated, and intelligent technology (such as neural networks, expert systems, reinforcement learning, etc.) needs to be used for processing, including the analysis and decision of the compaction process, the recognition of the compactability of the fill material, etc., which are still under in-depth study.

Below, based on the compaction data analysis, some commonly used rolling plans are presented. However, not all of this is true, and it is still necessary to accumulate experience (data) to enrich and improve the knowledge base of the control system continuously.

(1) **Rolling plan under normal conditions**

Under normal conditions (good fill material), determining the rolling plan is very simple. If the specified requirements (degree of compaction, compaction stability, control standards of compaction uniformity) are not met, continue to roll until the requirements are met, as shown in Fig. 4.12.

Figure 4.12 presents the rolling plan for compaction degree and stability (compaction uniformity will be discussed later). According to the analysis results, the current $E_i < [E]$, $\delta > [\delta]$, the given rolling plan is—to continue to carry out the rolling operation with the original process until the requirements are met.

(2) **Rolling plan under complex conditions**

Due to the complexity of the fill material and the diversity of the roller types and process parameters (i.e., the compaction), in many cases, it is necessary to conduct a comprehensive analysis and make decisions based on the compaction data. It cannot be judged solely on the compaction data of a single pass. Here are a few typical situations.

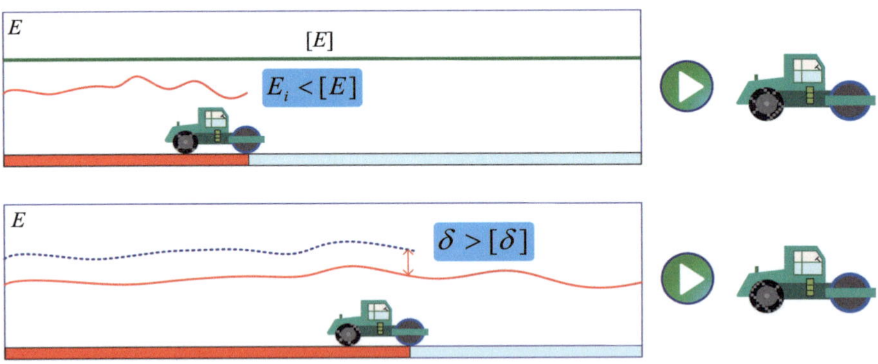

Fig. 4.12 Decision of the next action based on the analysis results

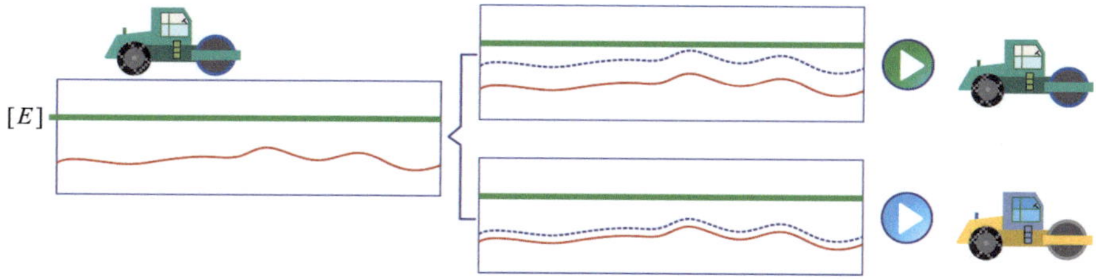

Fig. 4.13 Rolling plan A: change the compaction process based on compaction stability

In Fig. 4.13, if the compaction data does not reach the target value, continue rolling according to the original process. Still, sometimes you find that the rolling effect is not obvious—the difference in compaction data between the two passes is very small, indicating that the rolling under this process has reached a stable state, and the "compaction power" of the roller has been fully exerted. At this time, the compaction process should be changed (e.g., replacing the roller) and continue rolling until the target value is reached.

Adjusting the compaction process based on compaction data is a problem that requires in-depth study. It serves as the basis of intelligent rollers and is currently in the data accumulation stage. Adjusting the compaction process is also a key focus for the intelligent model and its learning, and reinforcement learning may play a significant role.

If the compaction data has reached the target value, there may also be two situations. One is that the difference in compaction data between the two passes is very small, indicating that it has indeed reached a stable state and the rolling operation should be stopped at this time; the other is that if the difference in compaction data between the two passes is large, indicating that the compaction is unstable, then regardless of whether the compaction data has reached the target value, rolling should continue until it is stable (indicating that the given target value is low), as shown in Fig. 4.14.

Another situation worth noting is shown in Fig. 4.15. Suppose the compaction data is below the target value. In that case, if you continue to roll, the compaction data of the next pass is lower than the previous one, indicating that the structure of the filler body has been damaged. If the difference between the two passes is significant, rolling

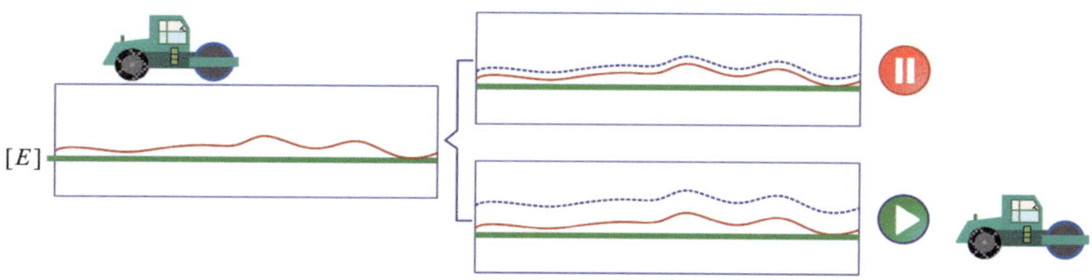

Fig. 4.14 Rolling plan B: decide whether to stop the rolling operation based on compaction stability

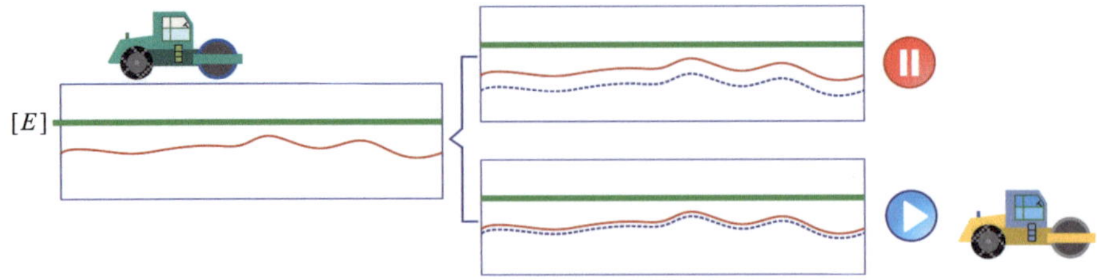

Fig. 4.15 Compaction scheme C: adjusting target values based on compaction stability

should be stopped. If the difference is small, you can try continuing with a "lighter" compaction process to see if there is a compaction effect (based on experience).

Based on practical experience, the main reason for the above phenomenon is that the target value is set too high and should be adjusted. Due to the complexity of the fill material, the target value (standard) determined by the test section may not be suitable, and it may need to be adjusted appropriately based on changes in the intelligent compaction data. This is also an important part of formulating the compaction scheme.

(3) About Compaction Uniformity

Under the same compaction process and the number of compactions, the compaction status of different areas will vary. This is a common phenomenon, primarily due to variability in the fill material, such as particle segregation and differences in moisture content distribution. When this difference is large, it will cause uneven compaction. When analyzing compaction data and making decisions, these factors should be considered.

As shown in Fig. 4.16, the analysis of the compaction data indicates that the data in the red area is lower than the specified control value, indicating an uneven area. The choice of compaction scheme should be used for the uneven area is worth further study. The practice has proved that simply increasing the number of compactions is not feasible in many cases.

According to engineering experience, for coarse-grained materials, unevenness is mostly caused by material segregation. In sticky, fine-grained materials, local high moisture content is a significant contributor to unevenness. It can be inferred that the fundamental reason for uneven compaction is the problem of fill material. Therefore,

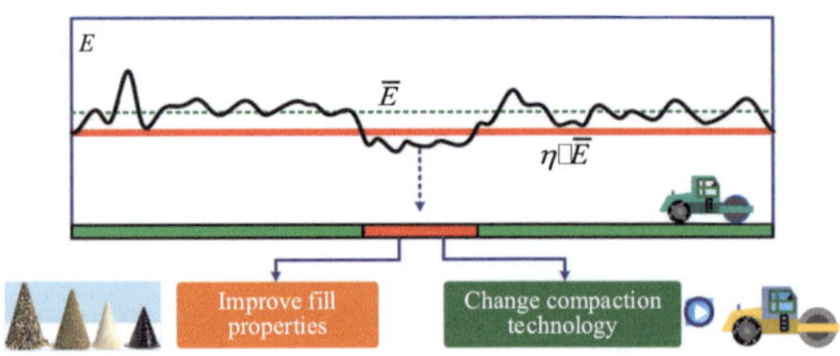

Fig. 4.16 Compaction scheme for compaction uniformity

improving the properties of the fill material and increasing its compactability are the primary tasks. Suppose you want to improve uniformity through compaction. In that case, you can try to change the compaction process, such as increasing the number of compactions, adjusting process parameters, or replacing the roller.

In summary, the control system (controller) can automatically determine a specific compaction scheme by analyzing and making decisions based on the compaction data and issue execution instructions to the actuator for implementation.

4.5 Execution of Compaction Work

If "analysis, decision-making" is "paper talk," then "execution" is practical work. For intelligent compaction, execution refers to the compaction work carried out according to the specified compaction scheme (instructions/information). The decision-making instructions include "stop compaction" and "continue compaction." Here, we discuss only the case of continuing compaction, focusing on the feedback control of compaction quality.

Compaction work requires feedback control. From the perspective of control, the chosen compaction scheme is a form of feedback information that serves as the basis for feedback control. The control content is divided into two parts: one is the control of the compaction process, including the roller type, excitation force, vibration frequency, and compaction speed; the other is the control of the fill material, including improving gradation, particle shape, and moisture content. These are the specific contents of intelligent compaction "execution," as shown in Fig. 4.17.

The concept of control has been introduced in Sect. 2.7. According to the feedback control principles outlined in Figs. 2.38 and 2.39, the working process of the execution stage (i.e., the working process of intelligent compaction) can be determined through the three control elements, as illustrated in Fig. 4.18.

The compaction degree, compaction stability, and compaction uniformity given in Fig. 4.18 are control indicators, each with its target value; the controlled objects

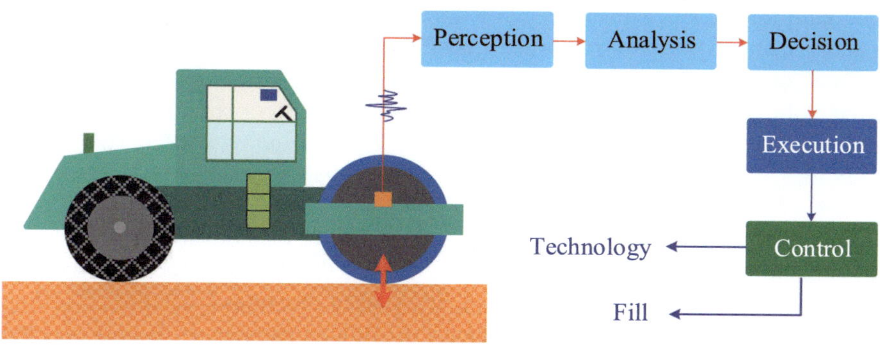

Fig. 4.17 Content of intelligent compaction execution

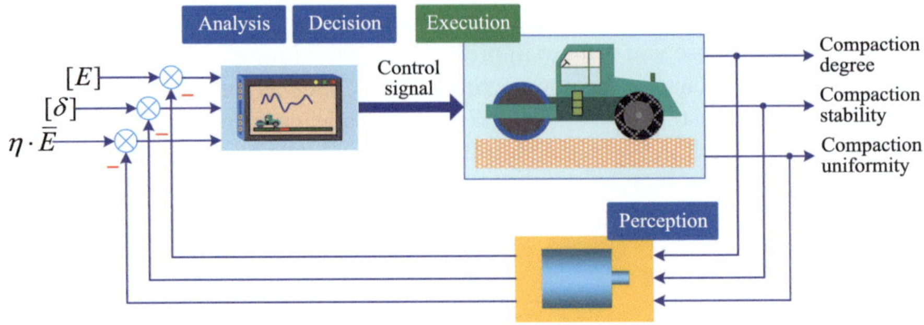

Fig. 4.18 Working process of the execution stage

are the fill material and the roller (vibrating drum); the controller generates control signals—mainly adjusting the roller process parameters (travel speed, vibration frequency, excitation force, etc.), and can also generate signals to adjust the properties of the fill material (particle shape, gradation, moisture content, etc.) (referred to as "soft controller").

The basic idea of feedback control is to compare the information obtained from analysis with the target value ($[E]$ and $[\delta]$, etc. in Fig. 4.18), determining the compaction scheme, and execute it based on the size of the deviation. By adjusting the relevant parameters, the deviation is minimized, thereby achieving the purpose of feedback control. In addition, this control (Fig. 4.18) is a multi-variable system, which can also be considered from the perspective of modern control technology. See the control technology volume of the series.

As mentioned earlier, although intelligent compaction has achieved automatic "perception, analysis, decision-making," the "execution" part still cannot be fully automated, and further efforts are needed to enhance intelligence. From a control perspective, there are currently three feedback control methods: manual, automatic, and intelligent, each with distinct characteristics.

(1) **Manual Control**

The main feature of manual control is human operation. During compaction, the operator manually adjusts the roller's operating parameters based on the decision information provided by the intelligent compaction control system, including rolling speed, rolling track, and vibration levels (weak, medium, or strong). The operator is part of the actuator; various controls for the filler, including particle size, shape, gradation, and moisture content, are all carried out either manually or via construction machinery.

Figure 4.19 shows a diagram of the manual control method, which is currently the most used compaction control method. During the rolling process, the control system automatically collects, analyses, and makes decisions, providing a rolling plan on the screen. This plan either adjusts the roller's parameters or improves the filler's properties.

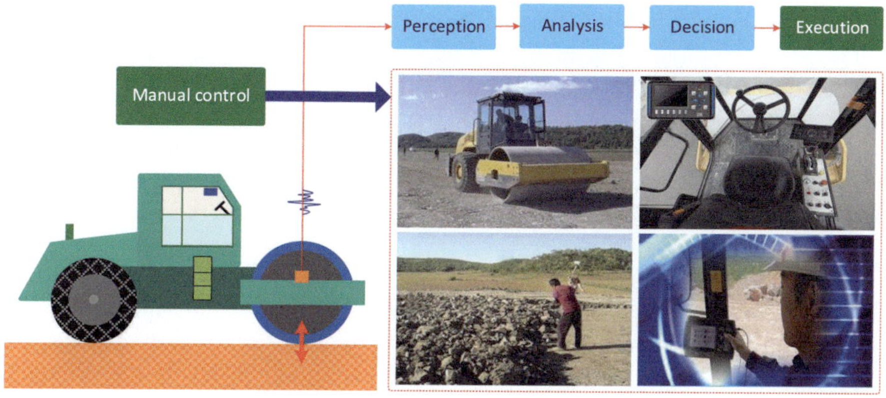

Fig. 4.19 Manual control method

The manual control method is applicable to ordinary vibratory rollers, and the process parameters that can be adjusted are limited. This means that the adjustment range of the roller's compaction force (the excitation force) is also limited and may not fully meet the compaction requirements.

(2) **Automatic control**

Automatic control primarily refers to the method of adjusting the roller's process parameters. It means that the roller can automatically adjust and control compaction process parameters (excitation force, frequency, speed, etc.) based on the feedback information from the control system via the actuator (mechanical or electronic device). It falls in the "object-object" type of control, that is, machine control of machinery.

Currently, the vibrating roller equipped with "automatic frequency modulation and amplitude modulation" capabilities can achieve this operation. When encountering a softer area, it will automatically increase the excitation force or amplitude, and when encountering a harder area, it will automatically decrease the excitation force or amplitude. The driver does not need to participate in the operation.

——*Automatic control is the mainstream in the field of control and the most widely used technology in industry. It generally refers to the use of additional devices to enable the machine, equipment, or production process (called the controlled object) to run automatically according to a predetermined rule without direct human intervention. A system that automatically controls the state of a controlled object is called an automatic control system, which is generally composed of a control device (including a controller, observer, and actuator) and the controlled object. It has three technical features: 1) It uses automated instruments (observers) to replace human sensory organs; 2) The controller (also known as the regulator) replaces the human brain; 3) The actuator (electronic mechanical device) replaces human hands and feet. Automatic control technology has a wide range of applications in intelligent construction, and the control technology volume of the series specifically introduces this knowledge.*

Given the current level of construction machinery manufacturing, it is not difficult to achieve automatic operation of construction machinery (including unmanned driving) through actuator control. The key issue is whether the control command is accurate! For the roller, the first core technology to be solved is to automatically and accurately perceive changes in compaction state. Currently, intelligent compaction has addressed this issue. However, there is another problem: how does the controller set and automatically adjust the compaction process parameters, and what is the adjustment range?

The relationship between the filler and compaction process is very complicated. The vibration compaction process to achieve the best compaction state of different fillers is not the same. Even for the same filler, the gradation is different, and the vibration compaction process to achieve the best compaction state is also different. However, the adjustment range of the roller's process parameters is now limited and cannot fully meet the demand. Automatically selecting optimal process parameter combinations for different fillers remains a challenge for the development of automatic rollers.

The filler still mainly involves manual processing or mechanical processing (manual operation). The key to achieving automatic control in this part is also the problem of correctly perceiving feedback control information. Regarding the filler, the coarse aggregate primarily exhibits gradation and segregation issues, while the fine aggregate faces problems with moisture content and distribution. The following gives a research idea for automatically improving the gradation and distribution of coarse aggregate, as shown in Fig. 4.20.

In Fig. 4.20, drones are used to "perceive" the particles and distribution of the filler, and then through the information platform, "cloud computing" is performed, allowing the machine to automatically "analyze" the image, obtain the gradation range and distribution status, and then compare with the required target (which needs to be determined in advance), make "decision-making" according to the difference, give a plan to improve the gradation and distribution, and issue "execution" instructions

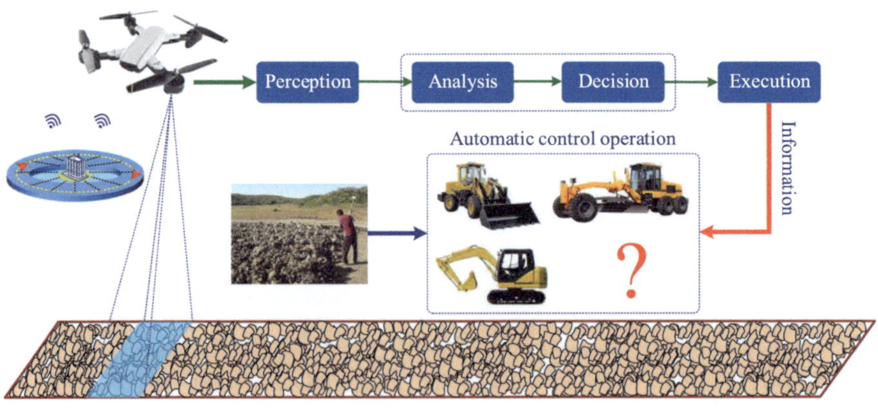

Fig. 4.20 The idea of automatically improving the filler

to the actuator (construction machinery). The construction machinery (as shown in Fig. 4.20, maybe new equipment needs to be developed?) performs the corresponding operation automatically according to the control command (information), such as improving the gradation by adding particles of different sizes and guiding the construction machinery to improve the distribution status of the filler on the paving surface. The current technology can fully realize unmanned and automated operations. The prerequisite is that all construction machinery should be equipped with information and networking capabilities, and implement microprocessors and software.

(3) **Intelligent control**

Intelligent control is the process by which intelligent machines autonomously achieve their goals. The current self-tuning control and adaptive control are low-level manifestations of intelligent control.

Intelligent control is developed based on automatic control, combining it with the advancement of artificial intelligence. Specifically applied to compaction quality control, it is advanced intelligent compaction technology, which is still under development. The current control system features automation and intelligent functions, primarily manifested in automatic data collection, processing, display, and storage, as well as intelligent computing. However, there is still a long way to go to truly intelligent compaction.

The above systematically introduces the steps to achieve intelligent compaction. It can be seen that intelligent compaction fully embodies the four basic characteristics of intelligent construction—"perception, analysis, decision-making, execution." Despite this, there is still considerable research space in mechanical models, hardware, software, and algorithms.

4.6 The Role of the Information Platform

After the compaction operation is completed, all compaction data is stored in the control system's computer. However, this type of computer is typically a single-chip microcomputer with limited processing capacity. It needs to import the data into a computer with greater processing power. The background management system should complete this part of the work to process the data further, generate various reports, and manage the data.

With the development of network technology, long-distance wireless data transmission has become increasingly convenient, and the demand for unified management of compressed data is also increasing. The single background management software has gradually evolved into a comprehensive management system that operates in a digital and networked manner. This is the concept of the information platform.

Currently, the compaction information platform has unified the transmission format of various on-site compaction data, ensuring guaranteed communication

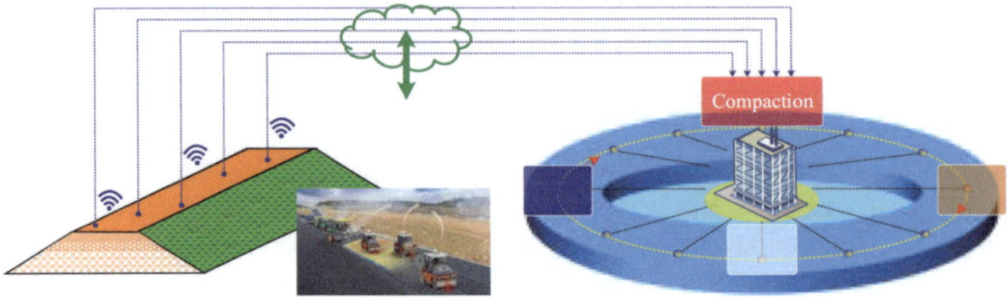

Fig. 4.21 Various information of intelligent compaction imported into the information platform

(Fig. 4.21), and is gradually moving towards standardization. It has evolved into a subsystem (or module) of the intelligent construction information platform system.

The China National Railway System and provincial transportation departments all have corresponding information management platforms, which include intelligent compaction data management modules. The US National Intelligent Compaction Standard Software Veta is also such software (it has now become the standard of FHWA and AASHTO). Additionally, some instrument manufacturers have developed data management software. Still, they are generally only suitable for the data transmission of their equipment without a unified data standard, which is not conducive to data sharing and comprehensive use.

The information platform serves as a bridge between various subsystems and users, acting as a mediator. All terminals communicate with the platform, which can uniformly process related information, and users can apply for various technical services through it. For example, you can view various compaction data on your mobile phone, submit for data processing (cloud computing) on the platform, and monitor the compaction process on-site in real-time, which is difficult to achieve with ordinary data management software.

With the help of the information platform, not only is the data format unified, but data sharing has also become possible, and remote cloud computing and real-time monitoring have been realized. This represents the embryonic form of the Internet of Things (IoT) engineering and serves as the foundation for the development of Cyber-Physical Systems (CPS) or digital twins. What is particularly important is that the information platform collects compaction data across various aspects, which is highly diverse and a kind of "engineering big data." It not only provides training data for machine learning but also serves as an important basis for establishing an intelligent compaction knowledge base.

Fig. 4.22 Filling material, roller, and compaction control system

4.7 Engineering Example

The primary purpose of developing intelligent compaction technology is for use in engineering applications. To further familiarize readers with the implementation process of intelligent compaction, an engineering application example is summarized based on many years of engineering practice for readers' reference.

The background of the engineering example is a high-speed railway subgrade construction site, with the filling material is "soil + a small amount of crushed stone," and an 18-ton vibratory roller for compaction, equipped with an intelligent compaction control system, as shown in Fig. 4.22.

In engineering applications, intelligent compaction is carried out in accordance with the construction sequence, and its basic characteristics, "perception, analysis, decision-making, execution," are all reflected in the specific implementation process. For users, what they perceive is a high degree of automation that requires minimal personnel involvement. The following outlines the specific application process in accordance with the construction sequence.

(1) **Preparation work**

From a mechanical point of view, vibratory compaction is a dynamic test (vibratory compaction test). It is essential to inspect the loading equipment (vibratory roller) and control system (mainly the measuring equipment) before the test.

As mentioned earlier, the fluctuation of the vibration frequency of the roller will cause a larger fluctuation in the exciting force ($P = me\omega 2$), leading to uneven compaction forces, which in turn cause uneven compaction. At the same time, as a loading device, if the applied load is unstable, the results of the dynamics test

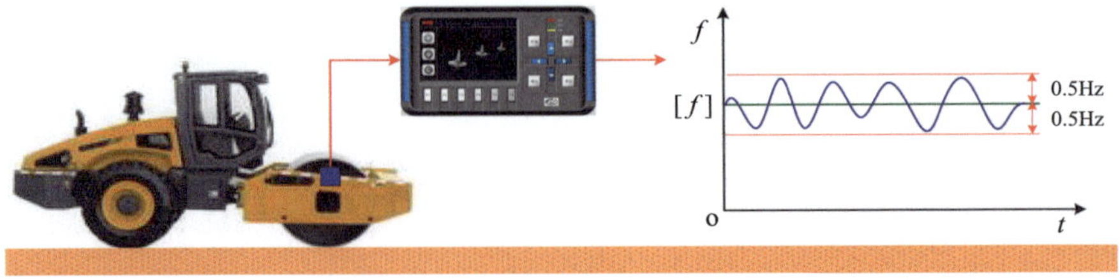

Fig. 4.23 Checking the fluctuation of the vibration frequency of the vibratory roller

will also be unreliable. Therefore, it is necessary to verify the roller's vibration performance and select a vibratory roller that meets the specified requirements as a continuous detection loading tool.

Due to the limitations of mechanical manufacturing, the roller's vibration frequency will fluctuate to some extent. The inspection of the roller's vibratory performance primarily tests whether the vibration frequency fluctuates and whether it deviates significantly from the rated value. Under the condition that the roller is moving at a normal rolling speed and vibration frequency, use measuring equipment (or other vibration testing instruments) to test the vibration of the vibrating drum and compare the actual measured frequency with the rated frequency $[f]$. It must meet the requirement that the difference be no greater than 0.5 Hz (as required by the Chinese railway and highway subgrade intelligent compaction technical standards), as shown in Fig. 4.23.

The inspection of the measuring equipment is simple: follow the relevant technical standards of the vibration testing instrument. Specific operations can be found in the literature [1].

(2) **Correlation verification**

Comparative experiments carry out correlation verification. The purpose is to verify the correlation between continuous control indicators (such as E_R) and point control indicators (such as K30), thereby determining the target value for intelligent compaction control. Comparative experiments are carried out in the test section. Figure 4.24 shows the rolling and comparative experiment plan of the test section.

The length of the test section should generally not be less than 50 m. According to the normal construction (compaction process is certain) requirements for filling, the test section is divided into three areas, each rolled into light, medium, and heavy compaction states, to obtain a larger range of compaction data and facilitate analysis of correlation and statistical relationships.

Regarding light, medium, and heavy compaction states, there is no standard distinction. The compaction degree of less than 70% belongs to the light state; the compaction degree of between 70 and 90% belongs to the medium state. The compaction degree greater than 90% belongs to the heavy state, among which the heavy area should reach the specified qualified standard to determine the target value. The rolling plan, as shown in Fig. 4.24, can be referred to.

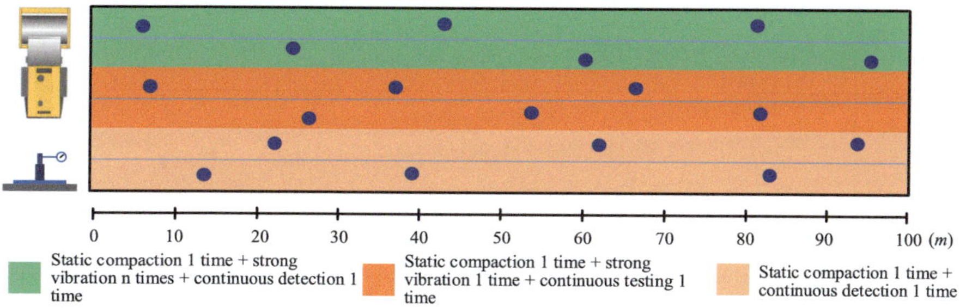

Fig. 4.24 Test section plan

The number of groups in comparative experiments is typically 9–18. Currently, the Chinese railway and highway standards specify 18 groups of comparative experiments, each comprising six experiments in light, medium, and heavy traffic areas.

Point detection is carried out using the plate load test method. The correlation coefficient determines the correlation between the point control index and the continuous control index. According to the standards of various countries, it is generally stipulated that the correlation coefficient must be greater than 0.7 to ensure statistical consistency between the two detection results.

This comparative experiment uses the subgrade reaction coefficient (K_{30}) as the point control index, and the continuous control index is the estimated modulus (E_R). The correlation coefficient is calculated using the least-squares method, and the correlation equation is determined by linear regression.

As shown in Fig. 4.25, the correlation coefficient is $R = 0.98$, indicating a strong correlation that meets regulatory requirements. According to the qualified value of point control $[K_{30}] = 130$ MPa/m, the target value $[E_R] = 150$ MPa is obtained through the regression equation, which serves as an important basis for process control of compaction quality in subsequent construction sections. In addition, the high-value area appearing in Fig. 4.25 is caused by the unevenness of the fill (more stones), and the phenomenon of coarse material concentration should be avoided in subsequent construction.

Next, the compaction quality control of a 100 m construction section is used as an example to see the control effect of intelligent compaction. The measuring system installed on the roller can automatically complete this process. The test section determines the fill and compaction process; however, there is a possibility of change during construction, which needs to be flexibly managed.

(3) **Rolling process control**

During the rolling process, continuous control of compaction degree, compaction stability, and uniformity are required. Among them, the compaction degree is controlled according to $[E_R] = 151$ MPa, and the pass rate is $[\beta] = 95\%$, the

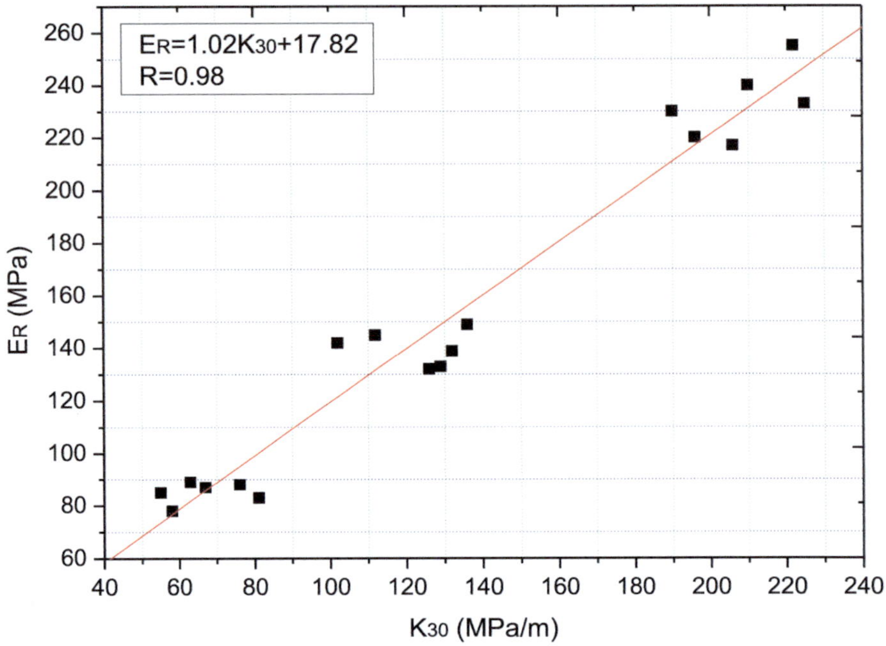

Fig. 4.25 Comparative experiment statistical results

compaction stability is controlled according to $[\delta] = 3\%$, and the compaction uniformity is controlled according to $0.80\overline{E}_R$. The control results are shown in Figs. 4.26, 4.27 and 4.28.

In Fig. 4.26, as the number of rolls increases, the compaction data approaches the target value $[E_R]$. From the second round, none of them reached the target value. In the 6th round, 95% of the area reached the target value, achieving control of the rolling process. The visual results are displayed in real-time on the cab screen, and compaction information for each point on the rolling surface is clearly displayed, facilitating quality control.

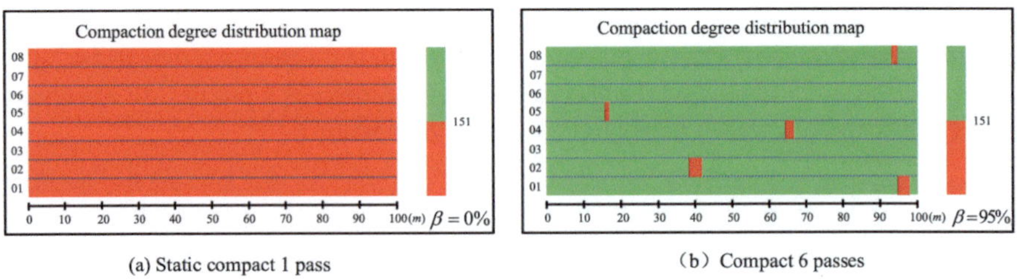

Fig. 4.26 Changes in compaction degree with the number of rolls

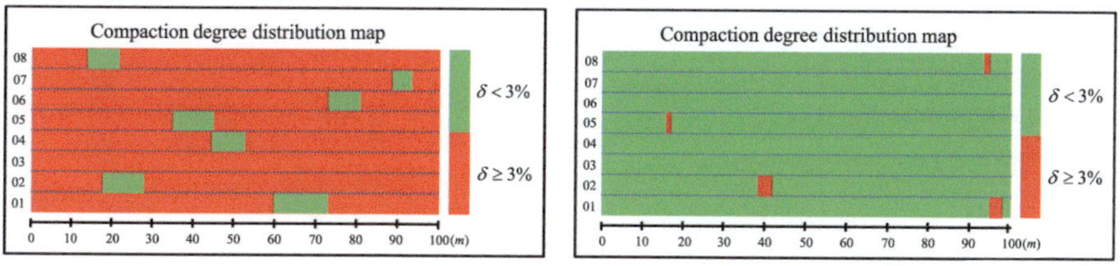

(a) Compact 5 passes (b) Compact 6 passes

Fig. 4.27 Changes in compaction stability with the number of rolling passes

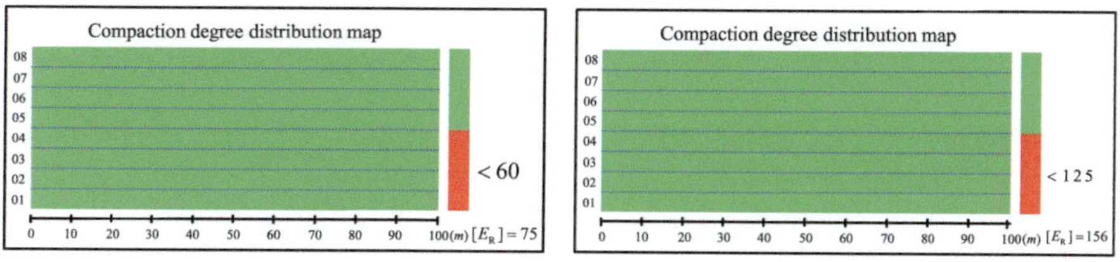

(a) Compact 2 passes (b) Compact 6 passes

Fig. 4.28 Changes in compaction uniformity with the number of rolling passes

Figure 4.27 shows the changes in compaction stability with the number of rolling passes. Due to the good nature of the fill, the relative difference δ between the compaction values of two adjacent passes gradually decreases. By the 6th pass, it basically meets the specified requirements (areas that do not met the standard can continue to be rolled, but if the area is very small, it can be allowed to remain).

Figure 4.28 shows the changes in compaction uniformity with the number of rolling passes. Although the average compaction value of each pass is not the same, no area less than 0.8 times the average value has appeared. This indicates a problem: under the stable conditions of the vibratory roller compaction process, the compaction uniformity mainly depends on the nature of the fill. It changes little with the number of rolling passes. Therefore, for uniformity, controlling the fill is key.

(4) **Quality inspection**

The quality inspection mentioned here includes continuous inspection of the rolling surface using a vibratory roller and point inspection using a plate load test. Continuous inspection can be carried out separately or using the last pass compaction data from process control. Point inspection mainly involves selecting inspection points, and the random selection method needs to be modified. According to the compaction status distribution map of the rolling surface, low-value areas are selected for point inspection.

With the completion of the continuous inspection, the rolling operation for this construction section is now over. The next task is to perform a point inspection using traditional methods. In accordance with the principle of cooperation between continuous control and point control, point inspections should be conducted in the

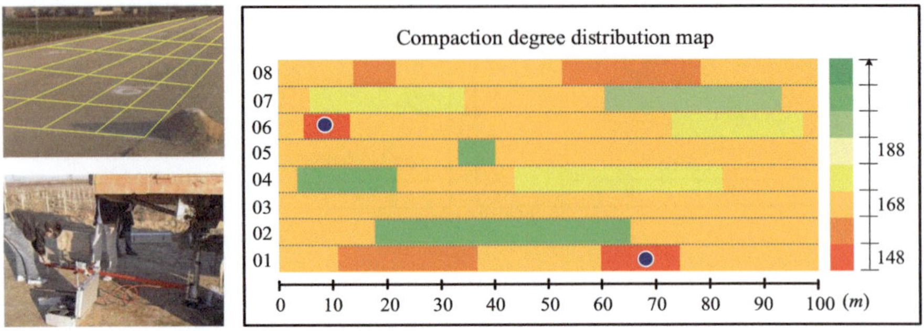

Fig. 4.29 Regular inspection of low points in weak compaction areas

weak compaction area (low-value area). Generally, at least two low-value points are selected in the weak compaction area for point inspection, as shown in Fig. 4.29.

Through the plate load test conducted on the 2 points in the weakest compaction area in Fig. 4.29, the obtained K_{30} values are all greater than $[K_{30}] = 130$ MPa/m. Therefore, the compaction quality of this construction section is judged to be qualified.

(5) **Compaction quality report**

According to regulations, the compaction quality report typically comprises relevant verification reports, compaction process archiving reports, and continuous compaction quality inspection reports. The information platform can automatically complete this part of the work.

The above introduces the entire process of applying intelligent compaction control technology, with a project example serving as the background. On the surface, this technology appears to be very convenient and easy to apply. But the actual situation is not the case. There will be various problems, such as the inability to roll to standard and multiple ineffective rolls, etc. These problems require the application of various new technologies, including intelligent technologies, to advance the development of intelligent compaction technology.

Chapter 5
Extended Application

Intelligent compaction technology is a product of multidisciplinary integration, encompassing field technology, mechanical theory, information technology, control technology, geotechnical technology, and mechanical technology, all of which have played crucial roles in the research, development, and application of intelligent compaction control systems. With the development of intelligent compaction technology, its application scope has also expanded significantly, and numerous related applications have emerged. Understanding these concepts can help illuminate the thinking behind research, development, and application.

5.1 Standardized Continuous Inspection

Intelligent compaction can continuously inspect the rolling surface, obtain the mechanical performance parameters (modulus) at each point, and determine the distribution of the mechanical parameters on the rolling surface. However, there is a problem with this continuous inspection: for the same fill, the inspection results from different rollers (or different process parameters on the same roller) are generally different. Although it does not affect quality control, it is not conducive to data circulation and comparison.

In various geotechnical material tests, the test results are often influenced by the loading equipment and loading methods used. For example, in the plate load test, different loading speeds and sizes (or different specifications of loading equipment) will lead to different test results, which is a manifestation of the complexity of the fill (Sect. 1.4). So, how does the plate load test solve this problem? The answer is standardization—unified regulations on loading equipment, loading methods, and data processing to eliminate these influences. In the plate load test procedure, clear regulations are established for loading equipment, measurement equipment, and

G. Xu et al., *Pioneer of Intelligent Construction - Intelligent Compaction*,
https://doi.org/10.1007/978-981-95-4184-3_5

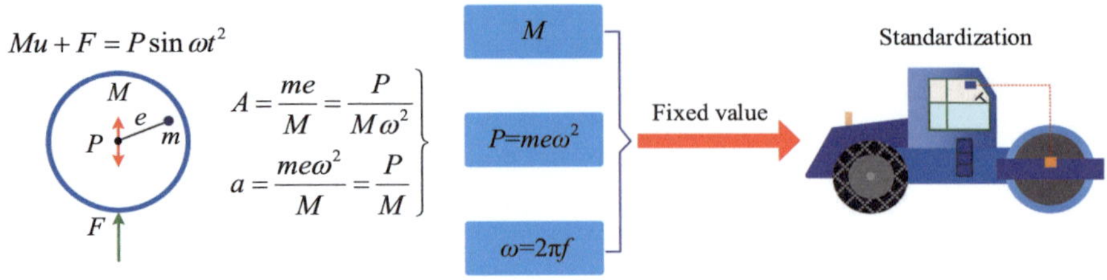

Fig. 5.1 Standardization of loading equipment

data processing, effectively achieving standardization. Inspired by this, continuous inspection of intelligent compaction can also be done in this way.

(1) **Standardization of content**

Standardization testing includes the loading equipment and testing methods. Intelligent compaction primarily involves standardizing the roller. According to vibration dynamics, the parameters that determine the strength of the vibrating drum are the vibration mass M, the excitation force P, and the vibration frequency f. As long as these three parameters remain unchanged, the nominal amplitude, A, and nominal acceleration, a, will also remain unchanged, and the vibration intensity is relatively fixed. Therefore, the so-called standardized continuous detection means maintaining these parameters at a constant level (strictly speaking, it should also include the walking speed), which is also the meaning of the standard roller, as shown in Fig. 5.1.

Intelligent compaction identifies the mechanical properties of the fill by continuously measuring the vibrating drum's responses. When the vibration mass is constant, the factors determining the size of the vibration response are the input—excitation force P and the resistance F of the fill, as shown in Fig. 5.2. Due to the complexity of the fill, if the excitation force P (size and frequency) changes, the detection results will differ, which is similar to the plate load test. Therefore, if you want to perform standardized continuous detection, you need to fix the combination of the roller's vibration mass (M), excitation force (P), and vibration frequency (f) to form the so-called standard roller, which standardizes the loading method and size.

Essentially, vibratory compaction can be viewed as a continuous dynamic test; however, compared to other mechanical testing methods (such as plate load tests and FWD), it has two shortcomings. First, the vibratory roller as a loading device, has a variable loading method, there is no unified regulation, and the loading accuracy is not high (with certain fluctuations); second, in terms of measurement, both force and

Fig. 5.2 Different rollers have different detection results

Fig. 5.3 Integrated detection robot

deformation cannot be directly measured, and indirect methods are used to obtain them, which is an indirect measurement method. Still, it is currently difficult to change the measurement mode. The problems in these two areas indicate that the current intelligent compaction technology is primarily used to control the compaction process. Standardized continuous detection is an extended application of intelligent compaction. Although its detection accuracy is slightly worse than that of the plate load test, the data obtained from all points in the fill are comprehensive, providing "full data," that has great practical significance in engineering.

(2) **Application**

Standardized continuous detection can be used as a standalone detection method or in conjunction with other detection technologies. The developing integrated detection robot combines continuous detection and steady-state vibration (fixed-point) detection, as illustrated in Fig. 5.3. Remote automated detection can also be performed via the Internet of Things.

This integrated detection robot can continuously detect the compaction surface, determine the compaction status distribution map, and then automatically perform steady-state vibration plate load tests in the weak compaction area, obtaining plastic deformation, deformation modulus, and elastic modulus, as shown in Fig. 5.4. Although both methods belong to dynamic testing, the former involves the interaction between the steel drum and the continuum, utilizing indirect measurement technology. In contrast, the latter involves the interaction between the bearing plate and the continuum, and employs direct measurement (force and deformation) technology. The latter's detection accuracy is higher than the former. The integration of the two can complement each other and make up for their respective shortcomings. In the process of technological research and development, it is also necessary to integrate multiple technologies (i.e., the so-called cross-border technologies).

The primary challenge in standardizing continuous detection is determining the roller's vibration mass, excitation force, and vibration frequency of the roller, which requires extensive engineering test data. The current practice involves formulating

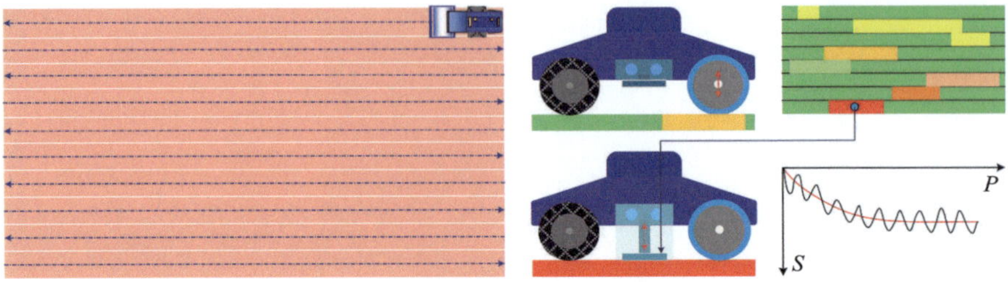

Fig. 5.4 Application of integrated detection robot

a set of vibration parameters based on engineering experience, establishing standards for them, and gradually refining this experience to transition to a unified standard.

5.2 The Key to Intelligent Roller

The current intelligent compaction can enable the roller to operate autonomously through "perception, analysis, decision-making." However, the "execution" part, given the roller's technical level, is still divided into manual control, automatic control, and intelligent control (see Sect. 4.5).

Manual control is the combination of a conventional roller and intelligent compaction, controlled by the driver's control of the roller's walking speed and compaction process parameters. This is currently the primary method of intelligent compaction control; however, the compaction process is limited, typically involving only two gears (or three to four), such as "high frequency with low amplitude" or "low frequency with high amplitude."

Automatic control is the combination of a road roller (automatic road roller) with automatic compaction process adjustment and intelligent compaction, in which the operator controls the roller's movement. At the same time, the machine automatically adjusts the compaction process parameters. Still, the adjustment range is unclear, and it is a trial-type adjustment.

Intelligent control, on the other hand, is based on automatic control, where the machine can intelligently adjust the compaction process parameters (with a clear adjustment range), which is a key technical feature of intelligent road rollers. The movement of the road roller can be controlled by the driver or by the machine itself (unmanned driving, see the next section). The development of intelligent road rollers with intelligent control as the core (primarily functional intelligence) should be a major direction for the future.

Intelligent road rollers should at least meet the following three conditions: first, they should have a rich compaction knowledge base, which allows the road roller (controller) to learn fully and master more skills; second, they should have a variety of adjustable compaction processes (or stepless adjustment), and the road roller

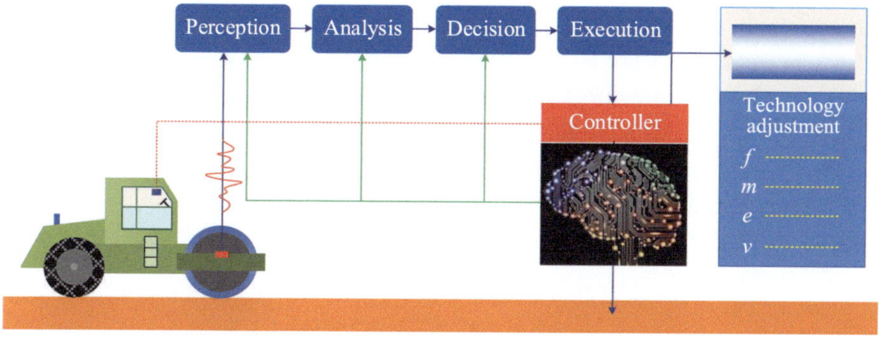

Fig. 5.5 Schematic diagram of intelligent adjustment of compaction process

can automatically adjust the change range of the compaction process according to the control command; third, they should be able to automatically identify the compactability of the fill material, propose control measures, and make the road roller smarter.

Among the three conditions above, the first is a necessary foundation, and the latter two are key. Figure 5.5 shows a schematic diagram of intelligent adjustment (control) of the compaction process (which also includes the identification and control of fill materials). The controller here serves as the "brain" of the intelligent road roller, requiring the ability to analyze compaction data and make informed decisions about the compaction process, and command the execution mechanism to adjust process parameters.

To equip the controller with the above capabilities, it must be trained, and a comprehensive knowledge base needs to be built (Sect. 2.6), both of which require a substantial amount of data. Similarly, the identification of the compactability of the fill material also requires practical experience, and its control measures cannot be fully automated at present; they can only propose control measures. The following focuses on the control problem of the compaction process, which is a significant difference between intelligent control and automatic control.

Intelligent road rollers intelligently and automatically adjust the compaction process parameters based on real-time feedback information to optimize the rolling operation. Currently, it involves two aspects: how to adjust the compaction process, and when to change it, and how much the process parameters need to be adjusted.

(1) **Ways to achieve multiple compaction processes**

For a road roller, the vibrating drum is the main body that generates vibration, the vibration mass M is constant, and the only adjustable parameters are the exciting force P, vibration frequency f (generally represented by the angular frequency ω, $\omega = 2\pi f$). Rolling speed v. Since $P = me\omega^2$, the real adjustable parameters are the eccentric mass m, eccentricity e, vibration frequency ω, and travel speed v, as shown in Fig. 5.6. Familiarity with these parameters is the basis for constructing multiple compaction processes.

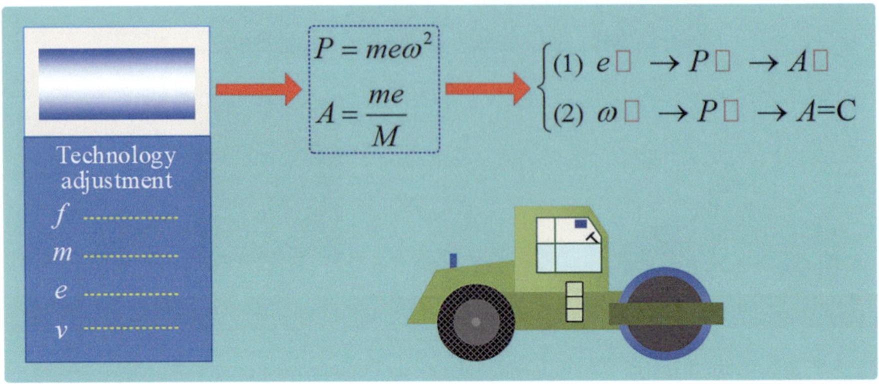

Fig. 5.6 Ways to achieve multiple compaction processes

For a moving road roller, the eccentric mass (m) is not easily adjusted automatically, and the parameters that can be automatically adjusted are mainly the eccentricity e, vibration frequency ω, and travel speed v. Based on changes in these parameters, a variety of compaction processes can be combined.

Figure 5.6 shows two basic schemes for adjusting process parameters. Scheme 1 is to adjust e, causing P and A to change, but ω remains unchanged; Scheme 2 is to adjust ω to change P, but A remains unchanged. Whether adjusting e or ω, from the current level of development of construction machinery, both are easy to achieve.

For Scheme 1, if the eccentricity e is increased, the exciting force P and amplitude A increase, and the vibration intensity becomes very large, making the road roller prone to bouncing (which will also affect mechanical performance). It is also easy to crush coarse-grained fillers. On the contrary, reducing eccentricity will also significantly reduce vibration intensity.

For scheme 2, increasing the frequency ω will only cause the excitation force P to increase ($\omega^2 \sim P$). Its vibration intensity is smaller than that of scheme 1. When the amplitude A remains unchanged, adjusting the process by varying ω may be more suitable for actual needs. If the frequency increases, the number of compaction forces per unit length will also increase, allowing the rolling speed to be appropriately increased (e.g., with a vibrating road roller used to roll the asphalt layer). Still, it is necessary to ensure the effective number of actions per unit length, see reference [1].

Combining the two basic adjustment schemes, you can first adjust the eccentricity e to change the amplitude A, and then adjust the frequency ω to make the excitation force P change from small to large, so that you can get different amplitudes, different excitation forces, and different vibration frequency compaction process combinations.

For directional vibration rollers, in addition to changing the size of the excitation force, you can also change the direction of action—from vertical vibration to horizontal vibration (oscillation) to adapt to different situations.

(2) The adjustment range of process parameters

The above provides a method for obtaining multiple compaction processes by adjusting e and ω. However, one problem is determining when to adjust the compaction process and how much adjustment is appropriate for e and ω. This is a problem worth studying. It needs to start with the interaction between the roller and the fill, and be solved using intelligent technology that leverages changes in information of the compaction state.

Regarding the question of when to change the compaction process, discussions have been made in Sects. 4.3 and 4.4, primarily based on the analysis and decision-making regarding changes in the compaction state, which can be used for reference. If the accumulated compaction knowledge is extensive (big data), an intelligent compaction expert system can be utilized for analysis and decision-making.

How does the roller clearly adjust the amplitude of the process parameters (e and ω)? According to current engineering practice experience, one approach is to accumulate experience (big data) to enable the roller to learn more effectively; the other is to use reinforcement learning to learn and accumulate experience during rolling. After a period of learning, the intelligent roller will determine the optimal control scheme autonomously, a key advantage of intelligent technology.

5.3 The Rise of Automated Rolling

With advances in science and technology, robotics has undergone significant development. In engineering, various construction robots, such as industrial robots, have emerged, which are essentially engineering machinery controlled by mechanical and electronic devices. The emerging autonomous driving roller is a construction robot (rolling robot) that automates rolling operations (part of engineering automation).

——*The so-called robot is a machine device that relies on its own power and control ability to perform certain tasks automatically through remote control or programming. This type of device has been used in industrial for a long time, and it has no direct connection to artificial intelligence (what is related is intelligent robots), primarily replacing human physical labor, such as welding robots, handling robots, and assembly robots. The current construction robots primarily replace manual operations performed by drivers. Still, many intelligent technologies have been integrated, and exhibit characteristics of "perception, analysis, decision-making, execution," although the specific content varies.*

The protagonist of automated rolling is the controller in the roller, which is the "brain" of the autonomous driving roller. It can "perceive" real-time geographic location information and surrounding environment information, conduct comprehensive "analysis" of location information and obstacle information, then make "decision-making" for the next action, issue "execution" instructions to the actuator, control the steering and speed of the roller, so that it can safely and reliably carry out automatic rolling operations in the planned site.

The principle of the autonomous driving roller is the same as an autonomous driving car, primarily relying on the car's computer system (controller) to achieve autonomous driving. However, the unmanned driving roller is technically simpler, mainly used for construction operations in a well-planned site, and the safety issue is less of a concern.

The perception technologies employed include satellite positioning, cameras, lidar, millimeter-wave radar, and infrared and ultrasonic radar. Analysis and decision-making primarily rely on AI technology, while execution mainly relies on automatic execution devices. Additionally, various other technologies are involved, including secure communication, Internet of Things, and remote monitoring.

For compaction operations, the above technologies are indispensable; however, a key technology is still missing: how to control compaction quality. Suppose the self-driving roller does not know the real-time change information of the compaction state of the fill. In that case, it can only control the compaction quality by the number of compactions (which is the current situation with automated compaction). Still, it isn't easy to truly control compaction quality. Therefore, without the intervention of intelligent compaction technology, automated compaction will not truly develop. Intelligent compaction is an indispensable technology, and requires integrating the intelligent compaction control system into the self-driving roller. Of course, the best solution is to integrate automated driving functions based on intelligent rollers and have them complete automated compaction operations.

Figure 5.5 shows a schematic diagram of the integration of intelligent compaction and self-driving rollers. The specific configuration method is simple: install an intelligent compaction control system on this roller. However, this roller needs to have the function of automatic frequency and amplitude adjustment, which is the automatic control mode mentioned earlier (see Fig. 2.40) (Fig. 5.7).

Automated construction is the primary direction of engineering construction and a key aspect of engineering automation. For fill construction, the integration of intelligent compaction and self-driving rollers is an inevitable trend. Suppose ordinary self-driving robots primarily rely on sensing their location and the surrounding environment to operate. In that case, intelligent compaction enables it to perceive the

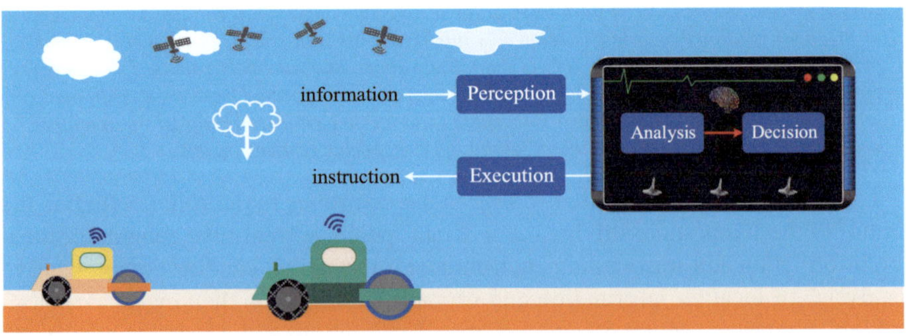

Fig. 5.7 Integration of self-driving roller and intelligent compaction

"underground" environment. It can monitor the compaction quality of the fill in real-time during compaction and make appropriate adjustments to compaction parameters (the current self-driving roller's compaction adjustment capability is limited). The practice has demonstrated that intelligent compaction technology is a crucial foundation for the development of automated compaction.

5.4 Cooperative Compaction Operation of the Fleet

Highways and railways are examples of strip-fill structures that stretch for hundreds of kilometers. For the construction of subgrade and base layers, a roller usually compacts the surface. However, for asphalt pavement, railway yards, and airports with larger work areas, multiple rollers are generally required to work on a compaction surface simultaneously, raising the issue of cooperative compaction operations among fleets.

When multiple engineering machines (rollers) work on a compaction surface simultaneously, a basic requirement is that there should be no collisions between the rollers (this is the same principle as self-driving cars). They should each follow the planned compaction route. From the perspective of cybernetics, this is a problem of cooperative control of multiple controlled objects (rollers) (see the control technology volume of the series). According to the characteristics of engineering construction, two control methods can be used to coordinate the compaction operations of the fleet.

The first method is decentralized control, where each roller has its own controller (a decentralized controller) that performs independent control. Still, there is communication between different decentralized controllers to coordinate the relative positions of different rollers. Decentralized control is the most commonly used control method in fleet operations.

The second method is hierarchical control, which is based on decentralized control and adds a central coordinator (a type of controller) that coordinates the relationships among various decentralized controllers (i.e., the relationships between rollers). As a result, decentralized controllers no longer require direct contact. For fleet operations, a roller with a continuous detection function can be added as a guide car (central coordinator) to uniformly coordinate the compaction operations across the fleet, including operating routes and compaction quality, as shown in Fig. 5.8.

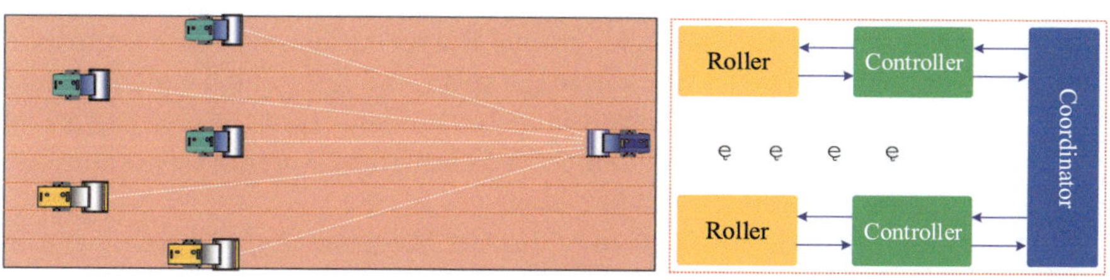

Fig. 5.8 Hierarchical control of roller fleet operations

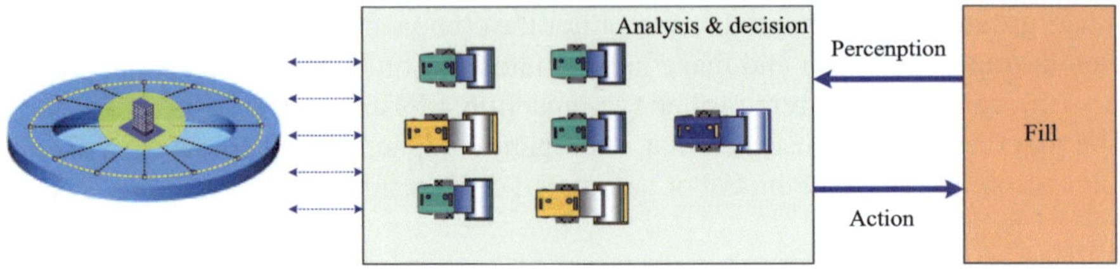

Fig. 5.9 Multi-agent system

Suppose the rollers in the fleet operation do not have an intelligent compaction control system. In that case, this guide car will transmit the detection results to each roller in real-time during continuous detection, and the rollers will perform compaction operations according to the corresponding instructions. Suppose the rollers are all equipped with a compaction control system. In that case, a standardized continuous detection car (the best option) can calibrate the detection results of other rollers, perform normalization, and achieve data sharing.

Fleet cooperative operations are a fascinating topic, and similar challenges exist across many industries. If it is a self-driving intelligent roller, it becomes a multi-agent system. This is a distributed intelligent group, and the primary challenge is coordinating its intelligent behavior based on perception information and collaborating effectively, as illustrated in Fig. 5.9.

When multiple intelligent road rollers are combined into a working group, they exhibit swarm-like behavior, requiring systematic behavior rather than independent action. At this time, multi-agent theory (which began in the 1990s) can be used to address these issues. Machine learning and expert systems will be applied. Additionally, reinforcement learning methods may be better suited to multi-agent problems, which were briefly introduced earlier.

5.5 Stratification of Filler Body

According to the theoretical knowledge introduced in Chap. 2, the modulus of the filler body, obtained from the vibration response of the road roller, is based on the theory of an elastic half-space and represents a comprehensive value over the test depth range. Due to the characteristics of the road roller's force, the vibration response not only contains information about the current fill layer but also includes information about the layers below, which differs from the plate load test (which generally only affects the current layer), as shown in Fig. 5.10. In many cases, it is necessary to separate the modulus of the current fill layer (E5) from the comprehensive value (E), which is a problem of parameter stratification.

The filler body is a multi-layer structure. In principle, layered elastic theory can be used for multi-layer modeling and processing. However, even the "forward problem" solution is numerical, and currently, only the multi-layer numerical solution under

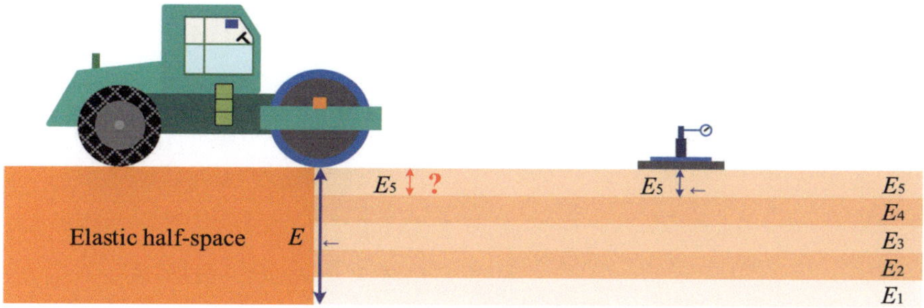

Fig. 5.10 Test depth of road roller and parameter stratification

circular load is available. The answer regarding steel drum action is not yet available, and the "inverse problem" of parameter identification is even more challenging, requiring alternative methods.

Given the characteristics of the road roller compaction operations, a two-layer elastic body model can be used to address this parameter stratification problem. The basic idea is shown in Fig. 5.11.

After the road roller continuously detects and obtains E_0, the filler body within this influence range is treated as an elastic half-space. Then another layer is filled (thickness h_1), and the measured modulus is $E_0^{(1)}$, this new elastic half-space body is the combination of the original elastic half-space body (E_0) and the current fill layer (E_1), that is:

$$E_0^{(1)} = f_1(E_0, E_1, h_1) \tag{5.1a}$$

This "current layer + half-space body" two-layer model is much easier to solve than the multilayer problem, especially for obtaining the modulus. If E_0, $E_0^{(1)}$ and h_1 are known, E_1 can be easily obtained. For specific calculations, please refer to relevant books on pavement mechanics.

Suppose you want to get the modulus of each layer. In that case, you need to start from the original ground base and carry out continuous detection layer by layer.

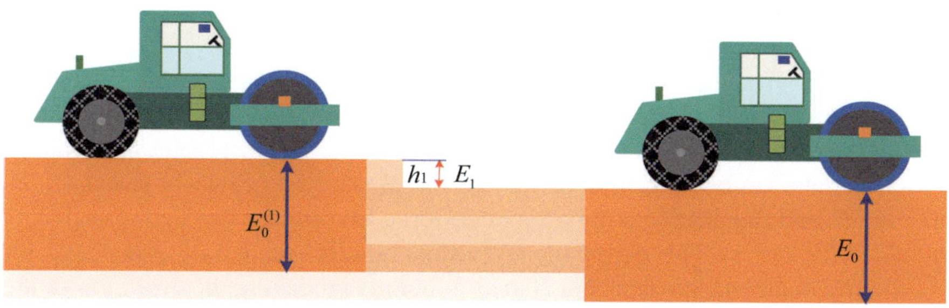

Fig. 5.11 "Current layer + half-space body" model

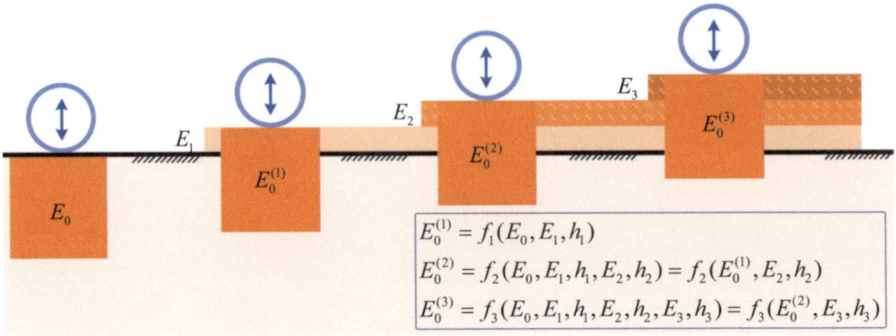

Fig. 5.12 Stratification of comprehensive modulus according to two-layer model

Then, solve using the two-layer model and obtain the moduli of all layers, as shown in Fig. 5.12.

Firstly, before filling, conduct continuous detection on the original ground (foundation) to obtain the comprehensive modulus E0, corresponding to an elastic half-space body. Secondly, after the first layer (thickness h1) is filled, repeat the continuous detection to obtain the new comprehensive modulus. $E_0^{(1)}$, corresponding to a new elastic half-space body, the difference from the previous elastic half-space body (E_0) is caused by the first layer (E_1). From this, the modulus of the first layer *E1 is* obtained according to the two-layer model, and so on, until all the moduli of the fill layers are obtained.

Separating the comprehensive modulus of the filler body obtained through continuous detection by the road roller into layers according to the fill layer is of practical significance. Firstly, it can establish a better correlation with traditional detection results and obtain better correlation; secondly, it can obtain the modulus of each point of the filler body, which is very useful for the mechanical analysis of the structure; finally, for the fill engineering, the modulus of each layer is the basic data needed for the digital model and virtual construction.

5.6 Digital Model of Filler Body

The data obtained at the end of construction is very rich, including modulus measurements at various points within the fill (obtained through continuous detection during intelligent compaction), material properties, gradation, and other relevant information. This data can be used to construct an initial state set primarily based on the modulus *E*, and to generate a digital model of the fill (see the Perception and Data volume of the series). This digital model (virtual fill) serves not only as the basis for formulating maintenance plans but also as a tool for predicting the evolution of the fill state, providing the database for virtual construction.

Since the virtual fill (digital model) stores the modulus data of each point within the physical fill if they are combined with the observed data within the real fill (obtained through embedded sensors, generally stress and strain), it is possible to analyze the state and changes of the entire fill based on the known initial modulus and changes in stress and strain at the observation points, as shown in Fig. 5.13.

Understanding changes in the state of the fill not only provides a basis for maintenance decisions but is also a key data requirement for virtual construction and is the main content of structural analysis (such as fatigue life). There are currently two methods for analyzing the state and changes of the fill.

The first is the state space method. First, establish the state equation of the system (road, railway) and predict and analyze its evolutionary behavior based on the initial conditions and state equation of the fill (system). The observed data can be used to calibrate the parameters in the state equation. This method is often used in modern control technology (see the Control Technology volume of the series). For road and railway engineering, the key is to establish the state equation.

The second is the intelligent method. When it is difficult to establish a mathematical model, an intelligent model can be considered for analysis. Given the large amount of data is left at the end of construction, along with some long-term observation data from observation points, the use of an intelligent model is feasible. For example, in the ANN model, the inputs can include vehicle load, temperature, humidity, modulus, stress, and strain. The vehicle load is the design value (including size and frequency), temperature and humidity are measured values, the modulus is obtained by intelligent compaction, and stress and strain are calculated by the mechanical model. The actual stress σ and strain ε are predicted using the ANN model, and the actual observed stress and strain data can be utilized as training data, as shown in Fig. 5.14.

It should be noted that the stress and strain calculated by the mechanical calculation are used as inputs, considering the correction of the mechanical model. If you do not consider the two inputs of stress and strain, it is also possible for readers to study them.

Additionally, the ANN model presented here is a basic form and can be modified and refined in practice. For example, the number of load applications can also be used as an input for fatigue life prediction, but it may require more training data, and more comprehensive data types.

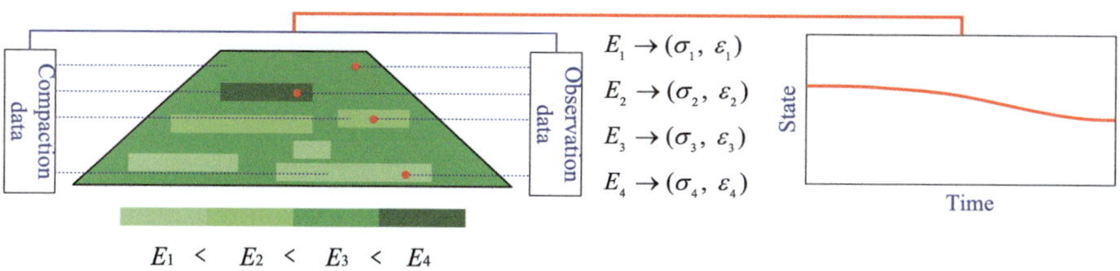

Fig. 5.13 Combination of digital model and test data

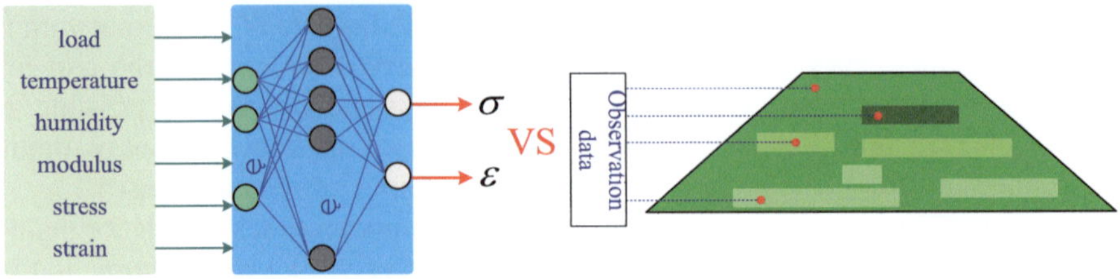

Fig. 5.14 ANN model for predicting stress and strain

A large body of practice has shown that the stress and strain calculated by the mechanical model do not match the observed stress and strain, and sometimes the difference is quite large. The reasons are multifaceted, with the mechanical model (the most mature of which is still the elastic model) being the primary reason. In this case, if there is a large amount of actual measurement data, it may be appropriate to use an intelligent model for processing provided it meets the engineering accuracy requirements. The application of intelligent technology in mechanics has a promising future (see the Dynamics volume of the series).

Intelligent compaction can determine the modulus of each point of the structure. Still, because the modulus of each point is different (non-uniform distribution), it is a function of position, which is (x, y, z),, which is different from the homogeneous body (modulus does not change), as shown in Fig. 5.15. Performing mechanical calculations and analyses on non-homogeneous bodies is a practical problem.

In general, the subgrade and pavement of roads, as well as the subgrade of railways, are analyzed using a layered elastic model. Regardless of how many layers the structure is divided into, each layer is a homogeneous body—there is only one modulus, which is one of the basic assumptions. If it is a variant body, it means that the modulus of each point (including Poisson's ratio and density) is not the same. It is a function of position, which will make it difficult to establish and solve the basic mechanical equation (including the constitutive equation). In this case, it is generally necessary to answer using anisotropic mechanics theory; however, the solution is more complicated and usually involves primarily numerical methods.

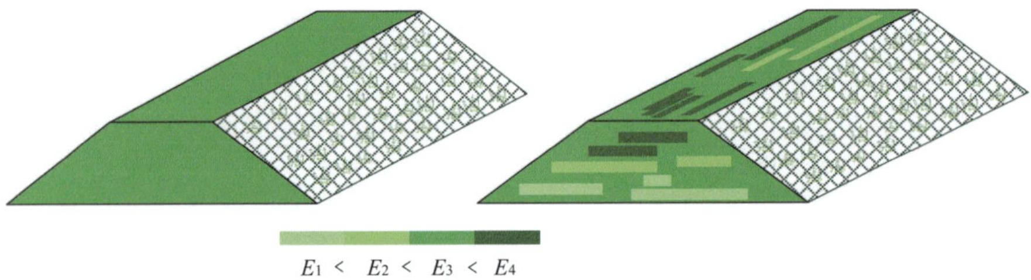

Fig. 5.15 Digital model of homogeneous body and variant body

5.7 Improving the Level of Virtual Construction

With the development of information technology, it is becoming more convenient to move things in the real world (physical world) into the computer, and it is becoming a hot spot of attention in all walks of life. This is virtual reality (VR)—using data from the real (physical) world, interactive three-dimensional modeling, and three-dimensional simulation on a computer to replicate the real world.

The so-called virtual is the creation of a simulation of the real world (digital model) on the computer, akin to "twin brothers" (the concept of digital twins), to enable simulation analysis of the manufacturing process and products. This kind of technology can be initially understood as an upgrade of past computer simulations (digital simulations). Still, the content is more comprehensive and richer, and the methods for building digital models differ.

Different names in various fields often refer to Virtual reality technology. Engineering construction involves virtual construction, where the engineering process is moved into the computer, enabling the virtually design and construction of roads and railways. This forms an interaction between the virtual and the real, as shown in Fig. 5.16.

In transportation infrastructure, virtual construction is still new. In the design stage, if the construction process can be virtually rehearsed on a computer, many potential problems can be identified and solutions found to avoid issues during actual construction. This simulation method is not new in other industries. It was later applied in the construction industry for various reasons, and insufficient data is one of the main reasons.

——As from the antient Chinese wisdom, *the stones of other mountains can be used to compete against jade, i.e., we can often learn something useful from other fields. Similarly, in the aviation field, engineers can design parts and assemble products on a computer, perform strength analysis, conduct aerodynamic calculations, program numerical control systems, conduct assembly inspections, and simulate motion, among other tasks. The manufacturing process is almost identical to that of real product. All parts can be described with a three-dimensional digital model, which can meet various needs and improve design efficiency and economic benefits. These successes are primarily due to a powerful database that stores*

Fig. 5.16 Moving engineering construction into the computer

extensive knowledge and data, including performance metrics for various components (such as fatigue characteristics). The virtual construction of transportation infrastructure should also follow a similar technical route. Accumulating data is crucial, particularly data on materials and processes. Continuous detection in intelligent compaction is a crucial means of obtaining performance data for the filler body.

Currently, in the field of road and railway engineering, virtual structures in computers are primarily represented as three-dimensional shapes, which constitute a type of digital model. Of course, some process management can also be simulated, but it isn't easy to simulate and analyze the performance of the construction process and the filler structure after forming. Therefore, the current virtual construction is still in its initial stage and requires technical improvement.

Virtual construction should be like other industries (such as aviation and automobiles), manufacturing products (filled structures) in the computer, and conducting various simulation analyses and calculations. Specifically, this involves selecting materials, construction machinery, and processes based on computer-generated design data, carrying out virtual construction operations (such as spreading and rolling), and analyzing and controlling construction quality to optimize the combination and use of materials, machinery, and processes. To find the best construction plan. For the virtually built roads and railways, it is possible to simulate and analyze the stress, strain (deformation), and life of the virtual filler body under the combined action of vehicle loads and natural environmental factors (temperature and humidity). Figure 5.17 illustrates some of the key aspects of virtual construction, but not all of it.

How do we implement the above ideas? The answer is data, specifically a variety of types of engineering data. We know that whether it is highways, railways, or airports, they are real buildings and exist in the real physical world. If they are moved into the virtual world, it is necessary to build digital models, which require substantial data support. However, the primary issue now is the scarcity of data types, particularly data on materials, structural performance, and processes. Intelligent compaction technology can address this issue and play a crucial role in advancing virtual construction.

Intelligent compaction can generate a large amount of engineering data, including the relationship among fill, process, and performance. This relationship can be

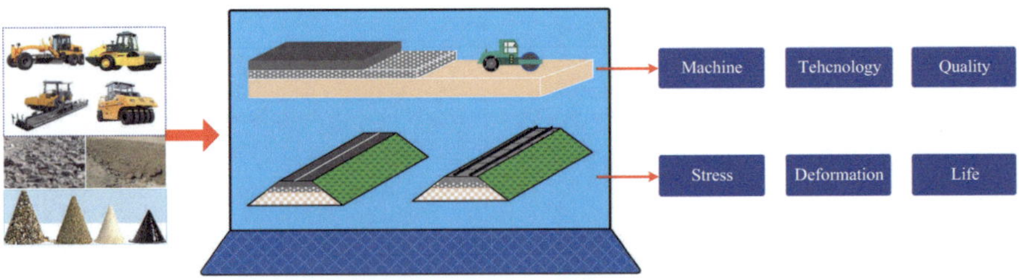

Fig. 5.17 Main content of virtual construction

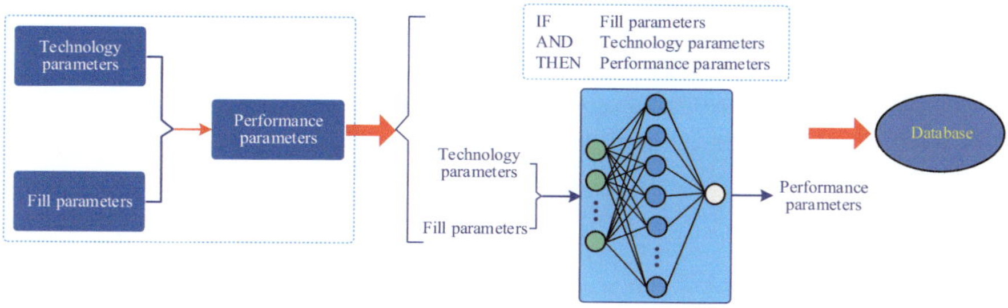

Fig. 5.18 Building a database

represented in the form of "IF–THEN" rules or by the ANN model, as shown in Fig. 5.18.

Suppose a large amount of this kind of data can be accumulated. In that case, a rich database can be built (with different performance data corresponding to different fillers and processes) so that the information in the database can be used to assemble, test, and analyze the structure according to professional knowledge and to support virtual construction.

For example, during the design stage, when virtually constructing a section of the subgrade, the properties of the filler (particle shape, gradation, moisture content, etc.) and the compaction process parameters (vibration mass, excitation force, frequency, rolling speed, etc.) must be specified first. The computer generates a specific thickness of the fill layer. Then, based on the correspondence between the filler, process, and performance (Fig. 5.18), various combination simulation analyses are conducted to identify the optimal solution.

For the simulation of the performance of the filler body, the modulus, Poisson's ratio, and density are required. Among them, the modulus can be obtained from continuous intelligent compaction detection results, whereas other methods must determine Poisson's ratio and density. For the physical parameters of the filler body, the stress, strain, deformation, and fatigue life of each point can be calculated under the vehicle load. This part primarily involves the mechanical theory of the interaction between the vehicles (such as a car, train, or airplane) and the surface body (highway, railway, airport, or city road).

Virtual construction not only helps optimize design schemes but also synchronizes with construction to build digital models. Through sensing and communication technologies, various information from the construction site is transmitted in real-time to the digital model (virtual body) on the intelligent construction information platform, allowing it to develop synchronously with the physical entity (filler body), as shown in Fig. 5.19.

Since the parameters in the digital model come from the real world, the virtual body (digital model) and the physical entity (filler body) have a very similar "twin brother" relationship so that various operations can be performed on the virtual body and feedback can be given to the real world, realizing the interaction between the

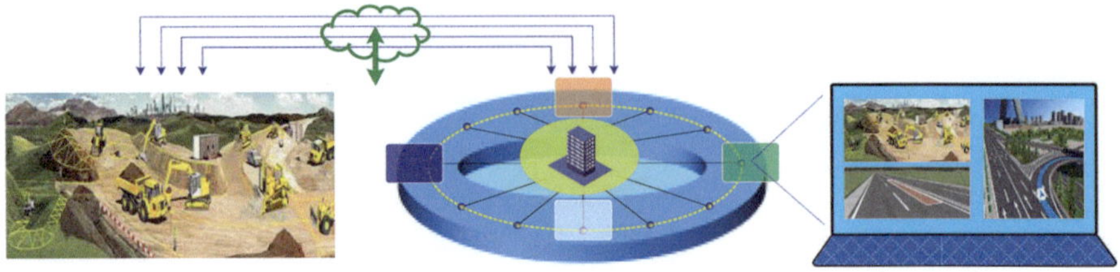

Fig. 5.19 Synchronous development of virtual construction and construction

two (Fig. 5.19). This is a form of Augmented Reality (AR) technology that can also be understood as part of Cyber-Physical Systems (CPS).

When construction is completed, a "twin" virtual filler body is generated during the formation of the fill structure, creating a virtual operational and maintenance environment. Suppose intelligent compaction data is combined with observation data and on-site detection data. In that case, it is possible to simulate the performance and lifespan of the structure (Sect. 5.6), and to conduct intelligent traffic simulation analysis.

At this point, the introduction to intelligent compaction has concluded. Some problems encountered in the application process will be answered in the Q&A section in the Appendix.

As a pioneer of intelligent construction, intelligent compaction has initially implemented autonomous "perception, analysis, decision-making, execution," but this is not enough. It should continue to develop towards advanced stages. There are still many topics worth in-depth research, such as basic theory, technical standards, control systems, and engineering applications. The integration of multiple disciplines should also be strengthened. During the above process, intelligent compaction will provide a great example for automated construction and advance the development of engineering automation.

Appendix A
Frequently Asked Questions (FAQ)

Given that this book does not elaborate on certain issues in the main text, the following will address frequently encountered and practical questions based on years of engineering practice experience, serving as a supplement to the main text.

1. **How to understand intelligent compaction?**

Intelligent compaction is based on continuous compaction control. Its basic principle is the same, but it is more intelligent, and the content is richer. To understand intelligent compaction, it is necessary to consider both mechanics and artificial intelligence (AI).

From the perspective of mechanics, intelligent compaction can be regarded as a mobile dynamic detection. However, compared with other mechanical detection methods (such as plate load tests and FWD), there are two shortcomings. One is that the vibratory roller as a loading device has various sizes of load (multiple excitation forces), and the loading accuracy is not enough (with fluctuations); the second is that in terms of measurement, both force and deformation cannot be directly measured, and indirect methods are used to obtain them, which is an indirect measurement method. These two problems indicate that the current intelligent compaction is primarily used for controlling the compaction process.

From the perspective of AI, the purpose of developing AI is to replace the human brain with a computer. According to the basic steps of human actions, the basic features of intelligent construction can be summarized as "perception, analysis, decision-making, execution." These four basic features are also the basic features and implementation steps of intelligent compaction, corresponding to four basic abilities, namely the ability to perceive compaction information, the ability to analyze compaction data, the ability to make compaction plan decisions, and the ability to execute compaction plans through rollers (mechanized or automated compaction operations).

In general, the current intelligent compaction is still in the early stages of development, and there are still many issues worth in-depth research. Despite this, this technology has become a strong driving force for the development of automated compaction.

G. Xu et al., *Pioneer of Intelligent Construction - Intelligent Compaction*,
https://doi.org/10.1007/978-981-95-4184-3

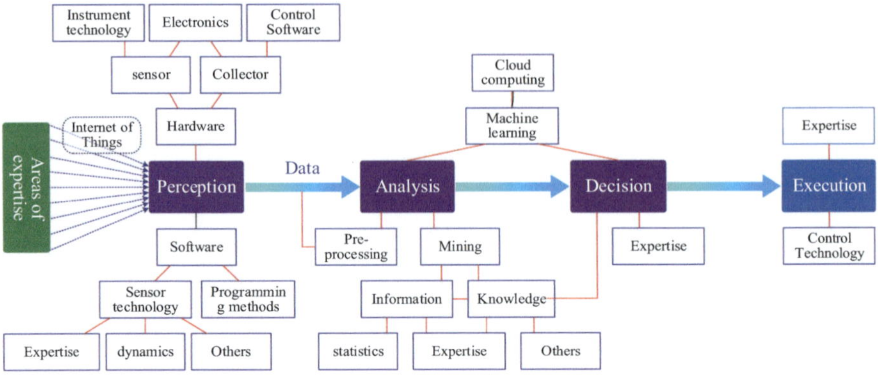

Fig. A.1 Sorting out the technologies involved in intelligent compaction

2. **What technologies are involved in intelligent compaction?**

Although intelligent compaction involves many disciplines, if you follow the "perception, analysis, decision-making, execution" route, you can sort out the various technologies it needs and their status, as shown in Fig. A.1.

In Fig. A.1, the hardware core of the "perception" part is sensors and data collectors (collectively referred to as perception terminals), which are common technologies. The software includes software for controlling data collection and software for identifying filler body performance parameters. The former is also a common technology, while the latter is a personalized technology that requires dynamics and professional technology (highways, railways, airports, urban roads, etc.). The "analysis and decision-making" part has both common technologies (machine learning algorithms) and personalized technologies (compaction quality analysis and compaction plan decision-making). The "execution" part mainly involves professional technologies and construction machinery technologies, which are specialized, while control technology is a general technology.

It should be noted that the technologies outlined in Fig. A1 are also involved in intelligent construction. These technologies are both systematic and to some degree independent, as detailed in the introduction to the series.

3. **How is intelligent compaction technology graded?**

Intelligent compaction originated from continuous compaction control in 1976. During its development, technologies and measurement devices based on different principles emerged. In summary, they can be divided into two major categories: empirical methods and mechanical methods. In order to sort out the characteristics of these technologies guide the technical research and development and engineering application of intelligent compaction, in 2017, the Federal Highway Administration (FHWA) of the United States issued an official document on the grading or classification of intelligent compaction technology (FHWA-HIF-17-046, the draft of this document was provided by the ISIC Executive Committee)—"The Development Direction of Intelligent Compaction Measurement Values (INTELLIGENT COMPACTION MEASUREMENT VALUES—A ROAD MAP)." In this technical document, the existing intelligent compaction is divided into five levels based on different control indicators. Levels 1–2 are primary technologies that rely on empirical methods, while levels 3–5 are advanced technologies that utilize the dynamic interaction principle between the vibrating roller and the filler body.

The document clearly states that the control indicators for intelligent compaction technology above level 3 should be mechanical quantities with clear physical meanings (such as modulus, stiffness system, resistance coefficient, etc.). The issuance of the document has guided the development and application of intelligent compaction technology. It has also accelerated the development of intelligent compaction technology in advanced directions.

4. **Is intelligent compaction applicable to the compaction control of asphalt pavement?**

The technical principle of intelligent compaction is the dynamic model of the interaction between the vibrating roller and the filler body. Here, the filler body is treated as an elastic body with no consideration of the effects of temperature and viscosity. Therefore, the current intelligent compaction is mainly applicable to the compaction control of subgrade and base layers.

Strictly speaking, if you want to use intelligent compaction in asphalt pavement, you need to re-establish the dynamic model, accounting for the effects of temperature and viscosity; however, the model will be more complex and difficult to solve. Even so, from practical experience, the current intelligent compaction is feasible on asphalt pavement, possibly for this reason—at high temperatures (such as above 80 °C), the temperature has little effect on the mechanical properties of asphalt mixtures.

When applying the existing intelligent compaction technology on asphalt pavement, two issues need to be noted. One is that the double steel drum vibrating roller cannot vibrate both drums at the same time and is only suitable for single drum vibration (from field surveys, it is generally single drum vibration); the other is that when the temperature of the asphalt mixture is below 80 °C, the modulus will significantly increase, but the density will not increase—that is, it cannot be compacted into a dense state (the reader can analyze the reason for this).

5. **What are the applicable conditions for the rolling pass control method?**

Although intelligent compaction technology above level 3 can correctly identify compaction quality, from market surveys, most engineering units still use level 1 technology for compaction control (the reasons are multifaceted). In many cases, because the control indicators of level 1 technology do not accurately reflect the compaction quality, these users primarily control the number of rolling passes (which can be easily recorded automatically through satellite positioning technology).

The rolling pass control method is an ancient empirical method; its usage conditions are very strict, and it is a method of last resort. Suppose the number of rolling

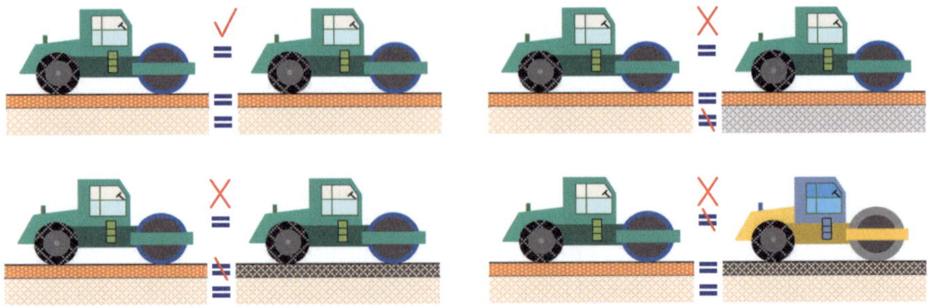

Fig. A.2 Applicable conditions for the rolling pass control method

passes obtained in the test section is applied to the subsequent construction sections. In that case, three conditions must be met—the same roller and process parameters, the same fill layer, and the same underlying layer (within the impact depth range), as shown in Fig. A.2.

In fact, the rolling pass control method is only applicable in cases where the fill material is uniform. For uneven fill materials, different numbers of rolling passes are required. That is, the number of rolling passes is not a fixed value. It needs to be optimized.

6. **What is the compactability of fill material?**

The compactability of fill material refers to its ability to quickly reach a compact state after a specific roller action. For coarse-grained materials, compactability is primarily influenced by particle shape and gradation; for fine-grained, viscous materials, compactability is mainly affected by moisture content.

In all construction specifications, regulations have been established regarding the compactability of fill materials, although the terminology used varies. For example, it requires good gradation of coarse-grained materials and maintains the moisture content of viscous fine-grained materials at the optimum level, among other requirements. The central idea expressed is to ensure the compactability of the fill material.

The compactability of the fill material is a key factor in determining the performance of the fill structure. If the compactability of the fill material is poor, regardless of the type of roller, compaction process, or control means used, it is difficult to ensure the compaction quality, and it is impossible to build a high-performance structure.

7. **How does intelligent compaction determine the compactability of fill material?**

Due to various reasons, in actual construction, the fill material does not always meet the specification requirements, and there will always be some fill materials that do not quite meet the requirements, such as inappropriate moisture content of viscous fine-grained materials, relatively smooth particle shape of coarse-grained materials, poor

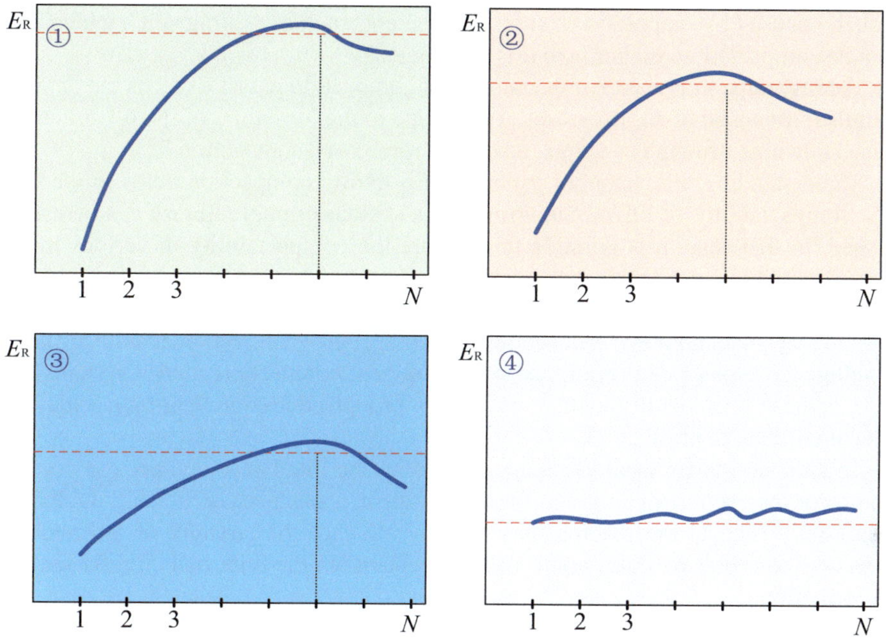

Fig. A.3 Schematic diagram of the relationship between intelligent compaction control indicators and filler compactability

gradation of fill materials, etc. This raises a series of problems, the most prominent being whether the fill material can be compacted.

By analyzing and summarizing a large amount of intelligent compaction application data, we found that the compactability of fillers can be indirectly inferred from the changes in the pattern of intelligent compaction control indicators (mechanical quantities, such as E_R). This discovery is particularly important for automated compaction and can guide the automatic compaction of road rollers.

For any filler where the intelligent compaction control indicator (E_R) changes well with the number of compaction passes (N), its compactability is also good (the correlation coefficient obtained from comparative experiments is also large), and vice versa, as shown in Fig. A.3.

Observing Fig. A.3, we can see that in ①, ②, and ③, E_R changes with N until it reaches its maximum value (which corresponds to the optimal number of compaction passes), but the magnitude of the change in E_R (slope) is different, reflecting the change in filler compactability. The change in E_R with N in ① is quite noticeable, indicating good filler compactability; the compactability in ② and ③ gradually weakens (the slope decreases), but the filler can still be compacted; in ④, E_R hardly changes with N, indicating that the filler is difficult to compact. For coarse-grained materials, situation ④ is usually due to unreasonable grading or relatively smooth particle shape, as confirmed through extensive practice. Sticky fine-grained materials are usually caused by inappropriate moisture content, such as "spring soil", which will not be compacted no matter how it is compacted.

Additionally, in Fig. A.3, if the number of compaction passes exceeds the optimal number, the value of E_R decreases. This is due to the loosening of the fill structure, also known as structural damage, resulting from "over-compaction."

Understanding the changing pattern of E_R during compaction helps to identify the compactability of fillers. Suppose such a standard sample library can be established. In that case, it is possible to compare the compactability of various fillers in a targeted manner, guiding the selection of materials, which is also the basis of advanced intelligent compaction technology. This method, known as pattern recognition in control technology, will be more advantageous when combined with artificial intelligence. Road rollers equipped with intelligent compaction control systems can master the skill of identifying filler properties through continuous learning, achieving true intelligent control.

In fact, no matter what mechanical quantity the intelligent compaction control indicator represents (modulus, stiffness coefficient, resistance coefficient), its change essentially reflects the macroscopic manifestation of the strength of the interaction between filler particles and reflects the change in the filler body's resistance to deformation.

8. **In what aspects is the complexity of the filler body performance mainly reflected?**

The complexity of filler body performance is primarily reflected in its non-linearity and time dependence. Non-linearity refers to the fact that the stress–strain relationship of the filler body is not linearly elastic, and there may also be plastic deformation. Different test methods and test loading methods for the same type of test generally yield different indicators. For example, using the same plate load test method, different modulus values (such as rebound modulus, primary deformation modulus, and secondary deformation modulus) will be obtained due to varying loading requirements.

Time dependence is reflected in the changing performance of the filler body. Typical examples include parameters such as subgrade modulus and moisture content, which change with the seasons. In addition, the results of the plate load test conducted immediately after compaction and those conducted after some time are often different (the latter is generally larger than the former), which is also a manifestation of time dependence.

The essence of the complexity of filler body performance lies in the geotechnical body. For embankment subgrades, their non-linearity is closely related to the compaction quality. The denser the structure, the better its linearity, but it is impossible to avoid time dependence completely.

9. **What is the relationship between the filler body performance indicators and the intelligent compaction control indicators?**

The performance of filler bodies needs to be considered from three aspects: strength, stiffness, and stability, which are also the three major indicators that all structures must meet. Although these performance indicators can clearly express the properties of the filler body, in practice, it is not easy to obtain these indicators, and they are generally difficult to obtain directly on-site. Therefore, people led to the introduction of multiple indicators that express the characteristics of the filler body, some of which are based on physical properties. Some are considered based on mechanical properties, such as compaction coefficient, porosity, rebound modulus, deformation modulus, foundation reaction coefficient, and bending settlement (vertical displacement), among others. Although there are many indicators, they all aim to characterize the stiffness, strength, and stability of the filler body.

In intelligent compaction, the control index is a mechanical quantity that represents the resistance to deformation of the filler body and corresponds to the stiffness of the filler body. Changes in the control index with the number of rolling passes (reflecting compaction stability) can indicate the stability of the filler body.

10. **Can intelligent compaction detect density?**

In the field of intelligent compaction, a common saying is, "Intelligent compaction can detect density." This is a wrong statement! This statement is equivalent to "a plate load test can detect density." Because intelligent compaction is a type of vibratory compaction test, it can be regarded as a moving plate load test.

Density and modulus are two important indicators for controlling the compaction quality of the filler body. The former is a physical indicator, and the latter is a mechanical indicator. They exist independently of each other (if they are theoretically related, they can be combined into one indicator). It is currently impossible to obtain physical indicators through mechanical tests using current technology.

Therefore, intelligent compaction cannot detect density; it can only detect the resistance deformation ability of the filler body, which is usually the modulus—based on the "estimated modulus (E_R)" of the roller. Still, the detection accuracy is lower than that of the plate load test.

Although density and modulus exist independently, in some cases, they show a statistical correlation, as numerous experiments have confirmed. Theoretically, it can be shown that, for general fine-grained materials (excluding the effects of water content and particle size), the change in density is approximately proportional to the plastic deformation rate. That is, the greater the plastic deformation of the filler body, the greater the dry density of the filler body, and the stronger the resistance to deformation. However, it should be noted that for cohesive fine-grained materials, there is an optimal water content problem; for coarse-grained materials, their resistance to deformation may continue to increase when the dry density stops increasing.

11. **Why control compaction degree, compaction uniformity, and compaction stability?**

This is mainly proposed for the structural characteristics of road engineering and railway engineering. It is essential to control these three aspects simultaneously during compaction, particularly for subgrade structures. The degree of compaction is the degree to which the physical and mechanical properties of the filler body reach the specified value, solving the problem of whether the filler body has sufficient strength and stiffness to support the upper structure; the uniformity of compaction is the degree of uniform distribution of the physical and mechanical properties of the filler body, solving the problem of whether it can uniformly support the upper structure; the stability of compaction is the degree of stability of the physical and mechanical properties of the filler body, solving the problem of whether the filler body can support the upper structure effectively and for a long time under the repeated load of driving (fatigue problem).

Of course, for other projects, it is also necessary to control the degree of compaction, uniformity of compaction, and stability of compaction simultaneously during the rolling process.

Additionally, when unevenness in the rolling is detected, the operation of the vibratory roller should be checked first, and then the filling material should be considered. In most cases, the unevenness of compaction is caused by the variation of the fill material. From the compaction stability analysis, the following information can be obtained. If the E_R does not change or changes slightly with the number of rolling passes N, there may be two situations. One is that the compaction power of the roller has been fully exerted, and there is no obvious effect on subsequent compaction. At this time, it is necessary to check whether the degree of compaction is qualified. The second situation is that there is a problem with the gradation of the fill material, and a good compaction effect has not been obtained.

12. **How does intelligent compaction control the compaction quality of fill stones?**

For fill bodies of the fill stone or aggregate type, traditional detection methods are no longer applicable. Suppose intelligent compaction is used to control the compaction quality. In that case, it is mainly used to control compaction stability. When using a heavy-duty vibratory roller for rolling operations, if compaction stability meets the accuracy requirements, the rolling can be stopped at that point. Additionally, continuous detection can also be carried out on the compacted sections to obtain the $E_{R.}$ The average value is the target value for subsequent similar construction sections. Of course, it may be necessary to adjust this target value according to the actual situation. In addition to compaction stability, compaction uniformity can also be controlled; however, it is necessary to adjust the level of variability, i.e., the correction factor η, as shown in Eq. (3.6).

13. **Why is there a need for related verification tests?**

There are three main reasons for conducting related verification tests. First, it is to support the application of new technology with a "crutch." When new technology is first used, people generally compare it with existing technology, and intelligent compaction technology is no exception. Although its control indicators are independent, traditional indicators (such as compaction degree and modulus) are deeply ingrained in people's minds, and they must be consistent to be accepted by the public. Secondly, the target value required for intelligent compaction control can be determined through related verification tests. Finally, because various detection methods and equipment use different technical principles, each has a different application range. Therefore, they must be distinguished, and related verification tests are the "sharp tool" to determine whether these methods and equipment are usable.

14. **How to conduct related verification tests?**

The purpose of related verification tests is to identify the connection between intelligent compaction control indicators and traditional point detection indicators (such as plate load test results), and to determine the target value of intelligent compaction control based on this connection. Therefore, comparative tests need to be conducted in the test section. Generally, the operation can be carried out according to the steps stipulated in the compaction standard. Additionally, the following points should be considered when conducting comparative tests.

① The test section should be compacted into three compaction states–light, medium, and heavy- and the heavy area must meet the requirements of conventional detection. Comparative tests cannot be conducted under the same state because they will cause sample points to cluster together, making it impossible for data to be spread out, and regularity will be masked.
② The correct order of comparative tests is to first conduct continuous detection on the test section and then conduct conventional detection, which can better reflect the actual situation.
③ Conventional detection should be completed as soon as possible to prevent changes in detection results due to excessive time, which can lead to timeliness issues and weaken the verification effect and the correlation.

15. **What factors affect the correlation between different detection results?**

Several factors influence the correlation between continuous detection and point detection. Firstly, several factors influence the acquisition of these two types of detection data. If the operation is improper, it will cause certain errors, and when combined, it will undoubtedly have a great impact on the correlation. Secondly, factors include test errors, detection equipment errors, differences between dynamic and static tests, and data processing methods.

The above-mentioned influencing factors exist when comparing different detection results (test results), which is a common phenomenon. The following focuses on the impact range involved in the roller detection and plate load test, which is the main influencing factor causing poor correlation of level 3 and above intelligent compaction technology, as shown in Fig. A.4.

The impact range of the roller's vibration load differ significantly from that of plate load test. First, in the vertical direction, the impact depth of the roller is several times greater than that of the plate load test. If there is obvious unevenness in the filling material within the impact range, the correspondence between these two detection results will become very weak. Secondly, in the horizontal direction, each data point of continuous detection represents the average value of the compaction state within a detection unit or footprint (the product of drum width and length). The measurement area of the plate load test is much smaller compared to the continuous detection unit.

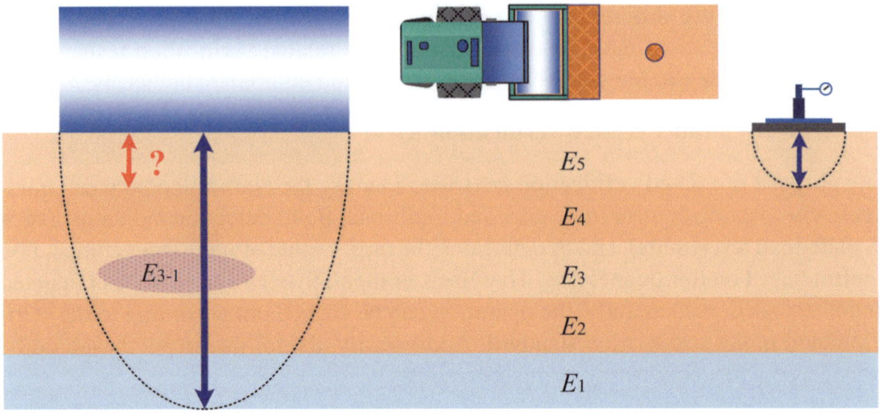

Fig. A.4 Schematic diagram of the impact range or influence depth of different detection methods

If the filling material is obviously uneven, conducting conventional tests at different points within a continuous detection unit will yield different results (as has been observed in the field), thereby worthening the correlation.

Based on the above two aspects of analysis, there may not necessarily be a strong correlation between the detection results (continuous detection) of level 3 and above intelligent compaction technology and the results of the plate load test (point detection) unless the filling material is very uniform.

For the 1–2 level intelligent compaction technology based on empirical methods, the correlation between its detection results and the results of the plate load test is generally poor. The main reason lies in the technical principles, as well as the above two influencing factors. The situation is more complex; therefore, it is recommended that the technology be upgraded to level 3 and above.

It should be noted, that for intelligent compaction technology at levels 3 and above, its correlation with the results of the plate load test is not required for normal application. Only when the properties of the filling material within the impact range are the same or similar will these two detection results have a good correlation.

16. **What is the relationship between the correlation coefficient and data dispersion?**

For a set of data (x_i, y_i), the index measuring the strength of their correlation is the correlation coefficient (r). $|r| = 0$ represents no relationship between the two. The data is scattered, $|r| = 1$ represents a very good correlation between the two. The data is not scattered, as shown in Fig. A.5.

Figure A.5 reflects the relationship between the correlation coefficient and the degree of data dispersion. As $|r|$ increases, the degree of data dispersion significantly decreases. When $|r| = 1$, the data is not scattered, which represents a deterministic relationship. In other cases, it represents a probabilistic relationship in the statistical sense, which must be clearly understood, as it involves understanding and using the correlation equation.

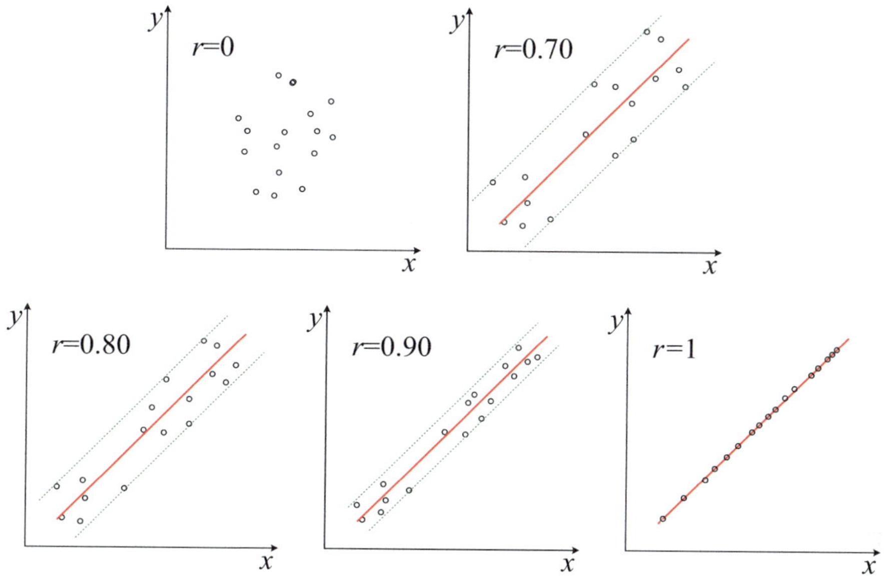

Fig. A.5 Relationship between the size of the correlation coefficient and the degree of data dispersion

17. **How to correctly understand and use the correlation equation?**

According to mathematical statistics theory, for a set of data (x_i, y_i), when the correlation is good (generally requires $|r| \geq 0.7$), a regression equation $y = a + bx$ can be established under the principle of least squares. This statistical equation is not the same concept as the deterministic equation we are accustomed to, and it cannot be used in the same manner as a deterministic equation. It needs to be correctly understood and used, which is often overlooked by many people. When using the correlation equation, the following points need to be noted.

① For the regression equation, the coefficients of its independent variable and dependent variable are determined by data statistics calculation and cannot be transformed and derived like a deterministic equation. For the regression equation $y = a + bx$, you can find y based on x, but you cannot find x based on this formula from y. You can only recalculate $x = c + dy$ Based on experimental data, we can then find x based on y. That is:

$$\text{Deterministic equation}: y = a + bx \quad \Leftrightarrow \quad x = (1/b)y - a/b$$

$$\text{Statistical equation}: y = a + bx \quad \times \Rightarrow \quad x = (1/b)y - a/b$$

In addition, people often ask the following question: If the regression equations between K_{30} and Evd and between ER and Evd are known, can the regression equation between ER and K30 be derived directly? The answer is naturally no, that is:

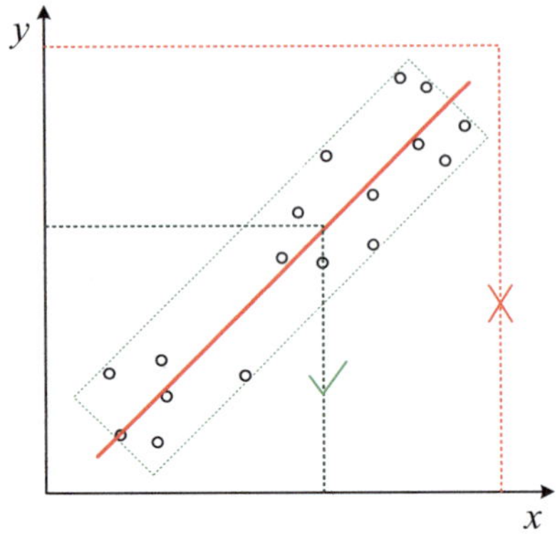

Fig. A.6 Interpolation and extrapolation of the regression equation

$$\left.\begin{array}{l} K_{30} = a + b \cdot E_{\mathrm{vd}} \\ E_{\mathrm{R}} = c + d \cdot E_{\mathrm{vd}} \end{array}\right\} \times \Rightarrow \quad E_{\mathrm{R}} = c + \frac{d}{b}(K_{30} - a)$$

Those who hold this view have made the mistake of treating the statistical equation as a deterministic equation. To determine the regression equation between E_{R} and K_{30}, there is no shortcut except to conduct experiments honestly.

② The calculation results obtained from the regression equation are not 100% reliable, but are a statistical result, which is related to the size of the correlation coefficient. Only when the correlation coefficient is 1 is it 100% reliable. Otherwise, certain errors will occur. According to practical experience, a correlation coefficient above 0.70 is considered strong, and the determined correlation equation has practical application value.

It is worth noting that determining whether a regression equation can be derived and whether it is effective are not the same thing. Regardless of the correlation coefficient, a regression equation can be determined based on experimental data; however, its suitability for use needs to be assessed based on the correlation coefficient. For example, when the correlation coefficient $r = 0.10$, a regression equation can also be obtained; however, since there is no correlation, it has no practical value.

③ The regression equation is generally only valid within the range of experimental data. Beyond the range of experimental data, although results can be calculated, the reliability of the results is not guaranteed. That is, data interpolation is possible, but there is a certain degree of dispersion, and data extrapolation is generally not possible, as shown in Fig. A.6.

18. **What is the critical point effect of the target value?**

In continuous control of compaction degree, such a phenomenon often occurs. When the target value $[E_{\mathrm{R}}]$ is reached, the degree of compaction meets the requirements.

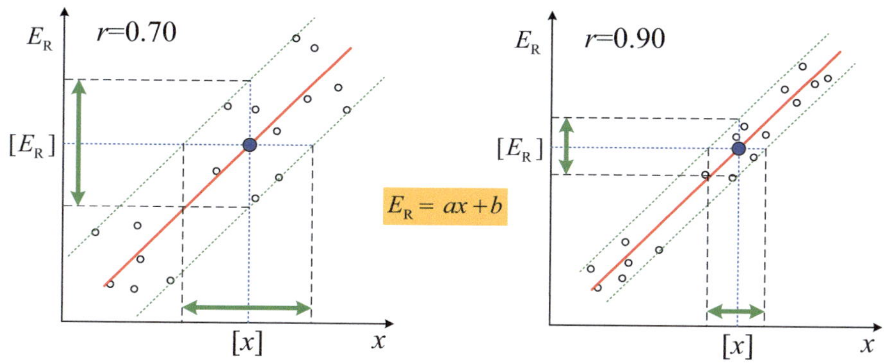

Fig. A.7 The discreteness of data determines the uncertainty of the target value

However, sometimes, it is not qualified to conduct traditional testing. We refer to this phenomenon of uncertainty that occurs near the target value as "boundary effects." For mathematical statistics problems, this is a common problem, but it troubles the field staff. Therefore, some analysis is necessary. There are two reasons for this phenomenon.

① The uncertainty caused by the discreteness of statistical data. As shown in Fig. A.7, the target value $[E_R]$ is obtained by substituting the qualified value $[x]$ of the conventional control index into the regression equation. It seems to be a one-to-one correspondence, but it is not. Because the regression equation is a statistical equation, there is no one-to-one correspondence at all.

According to Fig. A.7, the relationship between the target value $[E_R]$ and the conventional qualified value $[x]$ is constrained by the data dispersion width, and its range size is related to the correlation coefficient r. The larger the correlation coefficient, the smaller the fluctuation range and the smaller the uncertainty. However, if the correlation coefficient is not 1.0, there will always be a fluctuation range problem, resulting in uncertainty.

For example, according to Fig. A.8, $[K_{30}] = 130$ MPa/m corresponds to $[E_R] =$ 151 Mpa, but this is only a statistical correspondence. In actual continuous testing, $K_{30} = 130$ does not necessarily correspond to $E_R = 151$. $E_R = 151$ may also correspond to $K_{30} = 123$ (unqualified in traditional testing) or $K_{30} = 135$ (qualified in traditional testing), with some uncertainty. The linear relationship obtained by regression only has statistical significance and cannot be understood as a deterministic linear relationship.

② The uncertainty caused by the difference between the test section and the construction section. Suppose the properties of the test section and the construction section are not the same. In that case, it will also cause the phenomenon of E_R reaching the target value, but the conventional test is not qualified. This is because there is no completely identical section, and some variation in the fill

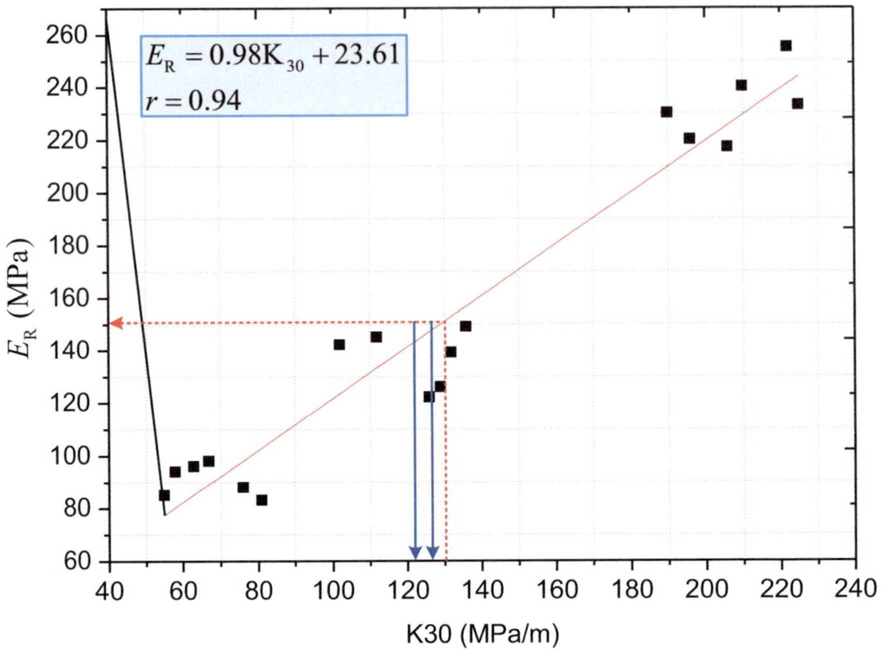

Fig. A.8 The correspondence between the measured target value and the conventional control value

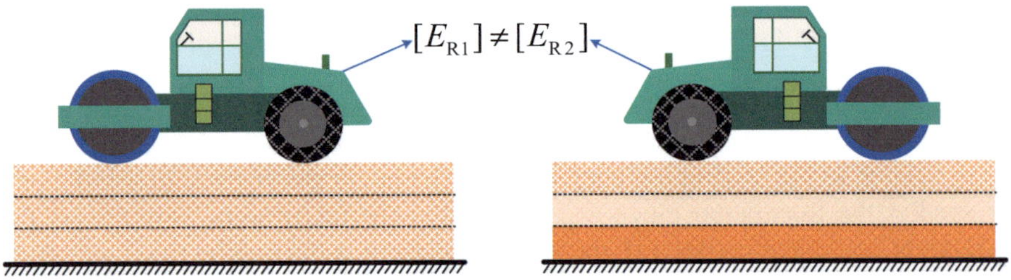

Fig. A.9 The difference between the target value of the test section and the target value of the construction section

occurs, which causes the predetermined target value [ER1] to differ from the actual target value [ER2], as shown in Fig. A.9.

When $[E_{R1}] \neq [E_{R2}]$, even if the measured $E_R = [E_{R1}]$, it cannot guarantee that the conventional test will pass. This uncertainty phenomenon is not primarily a boundary effect, but rather due to the variation in the filler. Of course, the boundary effect may also be present, making the situation even more complicated.

When the boundary effect occurs, the conventional test results are usually taken as the standard. But this does not mean that the conventional test must meet the requirements. It must be very correct. In fact, conventional testing also has its inherent

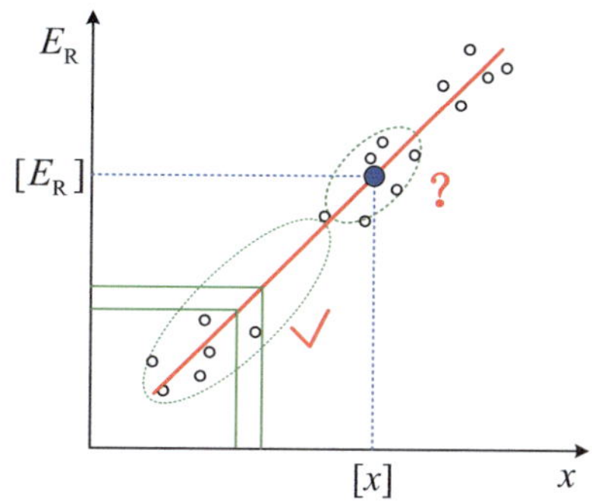

Fig. A.10 The two test results have consistency in the low compaction value area

defects. However, since the conventional test is the final inspection method specified in the current technical standards, it remains the primary judgment method.

It should be noted that the boundary effect only appears in the vicinity of the target value. When far from the target value, the results of continuous testing remain accurate. Therefore, except for a certain uncertainty in the vicinity of the target value, in other areas, especially in the low compaction value range, although there is a certain discrepancy between the results of continuous testing and conventional testing, the conclusion remains reliable, as shown in Fig. A.10.

The fundamental purpose of compaction quality control is still to prevent and control the occurrence of weak compaction areas, which is one of the sources of uneven settlement in the filler body, and this kind of weak compaction area is generally far from the target value area. Therefore, in this sense, continuous detection has its uniqueness. The critical point effect does not impact the control results of the "surface" and can still effectively control the weakly compacted area.

Regarding the problem of target value differences caused by the different properties of the test section and the construction section, in principle, the target value can be adjusted appropriately according to the actual situation to resolve it. Regarding adjustments, there is currently no unified model, and a reasonable choice must be made based on the actual site situation.

19. **Can the target value be applied to the same model of vibratory roller?**

From the perspective of intelligent compaction dynamics, if the compaction process parameters of the roller remain unchanged, the determined target value shows a certain universality. Therefore, for the same model of vibratory roller, the target value can be used interchangeably. However, due to differences in mechanical performance and the manufacturing dates, the vibration performance of the same model of vibratory roller may not be identical. The target values determined by different rollers may still differ, which need to be corrected according to the actual situation.

20. **Is the pass rate of the compaction degree of the rolling surface the qualification rate?**

In controlling the rolling process, the compaction pass rate represents the percentage of passes that achieves the intelligent compaction target value. Still, it is not the final pass rate for the quality inspection. Once the pass rate meets the specified requirements, the rolling operation can be stopped and a routine inspection can be requested. This not only optimizes the rolling process, avoids overcompaction and wastes working hours, but also reduces the risk of routine inspection failure.

21. **How should quality acceptance be carried out after using intelligent compaction?**

According to current technical standards, after the filler body is compacted, routine detection of sampling points should be conducted as the basis for quality acceptance. Due to the randomness of the sampling point selection, the inspection points may not necessarily be selected in the worst compaction state (even if selected, the distribution range cannot be determined), which may lead to "missed inspection." After using intelligent compaction, routine detection can be performed at fixed points on the compaction state distribution map to identify the weak compaction area.

When performing fixed-point detection, first perform it in the area with the lowest compaction state and a connected area of at least 5.0 m^2. If it is qualified, the entire rolling surface is qualified. Otherwise, point detection will be carried out in the next weak area with a connected area of at least 5.0 m^2, until a qualified area is detected. In this way, all areas that are unsuitable for engineering treatment can be identified. Therefore, the combination of continuous control and point control can each play to their strengths and jointly complete the task of quality control.

22. **What is the difference between the compaction state distribution map and the compaction degree distribution map?**

The compaction state distribution map is divided into several groups based on the relative size of ER on the rolling surface, within a certain data interval. Each group represents a compaction state, thereby dividing the rolling surface. The compaction degree distribution map is divided according to the target value [ER] on the rolling surface, into two zones: pass and fail, with a certain human factor influence (because the target value can be adjusted).

The compaction state distribution map is an objective map that accurately reflects the compaction status of each point on the rolling surface, without any human interference. It has no direct bearing on whether the rolling is qualified or not. The compaction degree distribution map is subjective. Its division is affected by human factors. Changing the target value can yield different compaction degree distribution maps.

23. **What changes will occur in the continuous detection results when encountering uneven underlying layers?**

The loading equipment for continuous detection is a vibratory roller, and its response reflects the filler body's resistance to deformation. Because of the impact depth of the vibratory roller is relatively large, some mechanical information in the underlying layer will be reflected in the test data. When the detection encounters a weak area in

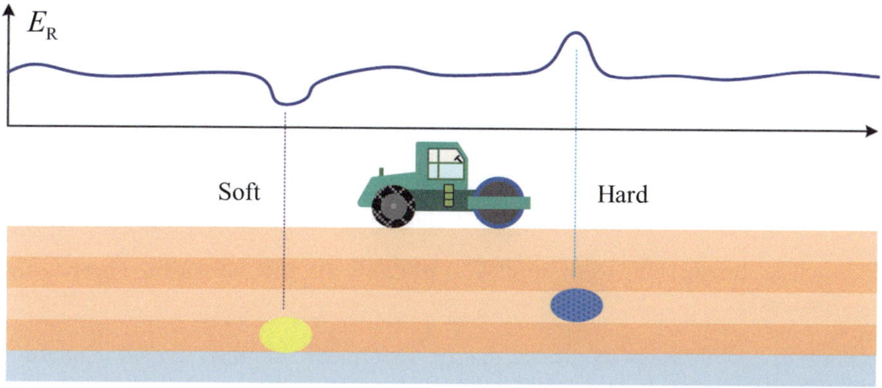

Fig. A1 Characteristics of vibratory compaction curve when the underlying layer contains different performance areas

the underlying layer, its E_R will decrease; when encountering a harder area, ER will increase, as shown in Fig. A.11.

In fact, the characteristic changes of the vibratory compaction curve are the same as the response of the fill layer itself, which contains uneven areas of soft and hard materials; there are only slight differences in the values. In addition, this phenomenon can also be explained from the perspective of wave dynamics.

24. **Can the number of routine inspections be reduced by using intelligent compaction?**

The decision on whether routine inspections are effective is not entirely a matter of the number of inspections, but more importantly, whether weak compaction areas can be detected. If targeted routine inspections can be carried out in areas where compaction is relatively weak, then the level of control will be greatly improved. Continuous detection in intelligent compaction can identify weak compaction areas, guiding the selection of points for routine inspections and thereby avoiding the uncertainty associated with random sampling. Therefore, in the long run, it is feasible and technically guaranteed to reduce the number of routine inspections after adopting intelligent compaction control technology.

Additionally, traditional point control primarily involves plate load tests and density checks. The purpose of the plate load test, in addition to the on-site evaluation of compaction quality, is to obtain the modulus of the fill, which is an essential parameter for mechanical calculations and analysis. From this perspective, even if intelligent compaction technology is fully adopted in the future, some point tests will still be carried out, as the modulus obtained from intelligent compaction remains "rougher" than that obtained from plate load tests. However, from the perspective of compaction quality control, if the performance parameters of future loading equipment, i.e., vibrating rollers, are uniform and stable (such as standard rollers), there is still a possibility of canceling point inspections (in fact, point inspections are not typically performed in coarse-grained materials). Because the current point sampling inspection control method also has major problems, especially in cases of uneven fill, whether more or fewer points are taken for inspection does not significantly affect the overall performance of the fill.

In summary, intelligent compaction focuses more on controlling the rolling process than quality acceptance inspection. Consequently, traditional point inspection methods must still need to be used for acceptance. Currently, routine inspections cannot be completely canceled; however, their frequency can be reduced.

25. **Is the linear vibration model applicable in the vibration compaction process?**

According to vibration dynamics theory, the linear vibration model (equation) assumes that the position of its equilibrium point is fixed and unchanging, and the vibration unfolds around this equilibrium point. However, the filler undergoes obvious plastic deformation during the rolling process, and the elevation of the rolling surface decreases constantly. There is no fixed point, and the damping coefficient and stiffness coefficient, which characterize its properties, are also changing. Therefore, in principle, the linear vibration model is not applicable to the rolling process of the fill. But after the rolling is complete, if the fill becomes an elastic body and meets the assumption of an elastic half-space, then this type of model can still be used for analysis. This conclusion also applies to the continuous dynamic model of the filler.

26. **Is the dynamic model established by the continuous testing mode approximate?**

The continuous testing model in dynamic methods is based on Newton's law, which, strictly speaking, belongs to the kinematic model. Whether there is an equilibrium point or not does not affect its force analysis and the establishment of dynamic equilibrium equations. Therefore, this model is not approximate but precise. It is an impact model.

However, since plastic deformation occurs in the fill during the rolling process, the expression of system resistance in this model is unknown, or it is very complex and difficult to solve directly; therefore, other methods must be sought for approximate solutions. Suppose the vibration displacement, speed, and acceleration responses of the vibrating roller can be measured at the same time, through certain technical means and practical experience parameters. In that case, relevant information about the fill's resistance to deformation can be obtained approximately, and the results of indoor and outdoor verification tests show that this is feasible. For media such as soil and rock, this approach can meet the accuracy requirements of quality control.

27. **Why does the measurement system of intelligent compaction need to be calibrated regularly? How to calibrate?**

The measurement system is an electronic device composed of various electronic components. Its performance will drift over time, affecting the accuracy of its output, so it needs to be calibrated regularly to eliminate this error.

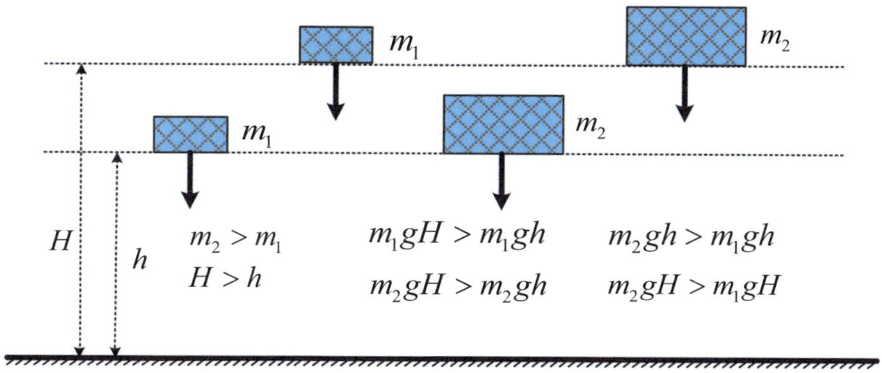

Fig. A.12 Simulation of compaction effect

The complete calibration work should be carried out in a measurement department with measurement qualifications. In regular testing work, users can often perform self-calibration tasks (generally every few months). For this, a high-precision vibration table needs to be prepared, and operation according to the user manual is sufficient.

28. **How do we understand "high frequency with low amplitude" and "low frequency with high amplitude" in the compaction process?**

For vibrating rollers, the primary factors determining the compaction effect are the size and frequency of the force (i.e., the number of times) and the distance of action (i.e., the amplitude). From a dynamic perspective, different mass blocks and varying falling heights are used to simulate the compaction effect, see Fig. A.12.

According to the theory of dynamics, the three elements that determine the impact of a mass block on the ground are the frequency of action, the magnitude of the force, and the falling height (or impact distance). Among them, the frequency of action determines the number of times the ground is impacted per unit of time. The higher the frequency of action, the more times the ground is impacted per unit of time, which is the same concept as the vibration frequency of a road roller; the size of the force determines the strength of the impact on the ground. At the same height, the larger the mass, the greater the impact on the ground, which is like the meaning of the exciting force of the road roller. The falling height determines the size of the energy when the mass block contacts the ground. The same mass block, the higher the height, the greater the energy of the impact on the ground, that is, $h = (1/2)mv^2$. The impact distance is equivalent to the concept of the amplitude of the road roller. The larger the amplitude, the stronger the effect on the compacted surface (under constant force), which can also be seen from the collision model.

According to the characteristics of these three elements, the greater the exciting force of the road roller, the higher the frequency, and the larger the amplitude, the stronger the effect on the fill. However, everything should have a "degree," and the extremes must be reversed. For the filler, excessive action can cause problems such as crushing and disintegration, thereby destroying the structure the filler is intended to provide. Therefore, the parameters of the vibration compaction process should match the properties of the fill. For the road roller itself, excessive operation may affect its mechanical performance and accelerate damage to the equipment. Therefore, when the vibrating road roller leaves the factory, it generally adopts the combination of "high frequency with low amplitude," "low frequency with high amplitude," similar to the combination of (m_1g, H), (m_2g, h) in Fig. A.12.

The above combination is based on the nominal amplitude A, which, according to $P = me\omega^2$, is given by $A = me/M = P/M\omega^2$ for reasonable matching. It should be noted that due to various reasons, such as mechanical damage, the parameter combination of “high frequency and high amplitude” is generally not used.

High frequency with low amplitude—when the vibration frequency of the steel drum (ω) is large, according to $P = me\omega^2$, the exciting force is already at a high value. If (me) is higher, the amplitude $A = me/M$ will be larger, which will lead to a larger exciting force, similar to the (m_2g, H) combination in Fig. A.12. Therefore, when the exciting frequency (ω) is large, it is necessary to adjust the value of the exciting force by reducing me. At this time, (me) is low, and the corresponding amplitude is also low, similar to the (m_2g, h) combination in Fig. A.12.

The low frequency with high amplitude—if the exciting frequency (ω) and (me) are both relatively low, it will lead to a lower exciting force $P = me\omega^2$, similar to the (m_1g, h) combination in Fig. A.12. Therefore, when the frequency (ω) is low, it is necessary to increase (me) for adjustment to maintain an appropriate energy output, similar to the (m_1g, H) combination in Fig. A.12.

29. **Is there a relationship between the rolling speed of the road roller and the vibration frequency?**

The vibration frequency of the road roller determines the number of actions on the filler body per unit of time, and there is a certain relationship between it and the rolling speed—the faster the rolling speed, the higher the vibration frequency should be to ensure the number of actions per unit rolling area. According to the empirical formula provided by the road roller manufacturer, there is an empirical relationship between the rolling speed of the road roller, v, and the vibration frequency, f, as shown in Fig. A.13.

The empirical formula given in the figure is the principle for selecting and controlling the rolling speed according to the vibration frequency of the road roller. For example, when the vibration frequency of the road roller $f = 30$ Hz, its rolling speed should not exceed 3.24 km/h.

30. **Is the compaction effect better the larger the tonnage of the vibrating road roller?**

The answer is no! The compaction effect of a large-tonnage road roller is not always beneficial, and it should be considered in conjunction with the type of fill material and the thickness of the fill layer. Currently, whether in a highway or a railway subgrade, the fill layer thickness generally does not exceed 40 cm. For fine materials,

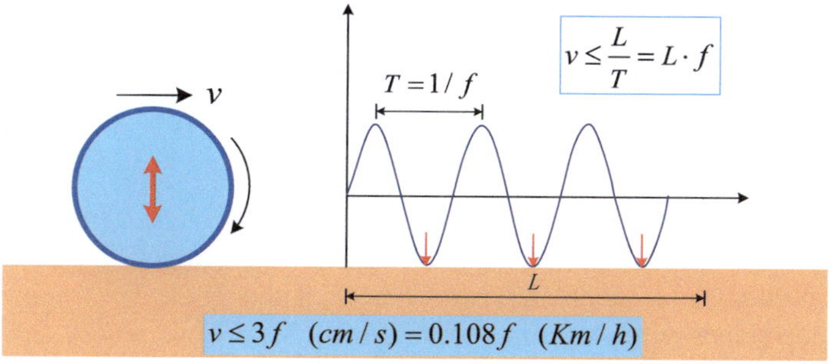

Fig. A.13 Relationship between the travel speed and vibration frequency of the road roller

a 14–16 ton vibrating road roller can generally meet the requirements; for coarse-grained materials, an 18-ton vibrating road roller can generally provide sufficient compaction. In addition, under vibration compaction, the fill particles experience "positioning" issue due to the vibration load and need to be repeatedly adjusted to find their suitable position. If the tonnage of the road roller is too large, it will force the particles to change their position, but it may not be the best position. It will also crush the particles themselves, increasing their crushing rate and destroying the original particle gradation. It may not be beneficial to form a high-quality structure. For stone-filled subgrades, large-tonnage road rollers are necessary for effective compaction; otherwise, achieving a compacted surface is difficult. This is because the fill thickness of the stone-filled subgrade is greater than that of general fill materials, and the contact area between particles increases, resulting in an increase in friction. The force needs to be overcome during rolling to allow the particles to move relative to each other and find their suitable packing positions.

Generally speaking, there is a certain relationship between the tonnage of the road roller, the particle size of the fill material, and the thickness of the fill. The larger the particle size and the thicker the fill layer, the greater the tonnage required for the road roller.

31. **What development stage has intelligent compaction reached?**

Intelligent compaction (IC) was developed in the 1970s, building on the concept of continuous compaction control. Over the years, IC has evolved to include more dynamic methods as well as intelligent and automated technologies. This advancement is particularly evident in the four key areas of "perception, analysis, decision, and execution" (see Fig. 1.17). A brief explanation of these areas is provided below.

(1) **In terms of "Perception"**

The goal of intelligent compaction is to utilize the vibration response of the roller during the rolling process to effectively control compaction quality. A key priority in this area is understanding how to perceive the quality information of fill materials (the constructed layers) from their vibration responses. Significant advancements have been made in intelligent compaction in this regard, particularly in the selection of control indicators and the adaptable use of testing technology.

In selecting control indicators, it is now common practice to view the rolling process as an interaction between the roller and the compacted materials, and to use dynamic principles for modeling this interaction. The control indicators arising from this perspective are mechanical quantities with clear physical meanings, which align with the technical characteristics of vibration compaction. For more detailed information on these concepts, refer to Chap. 2 of this book. For a more comprehensive understanding, please refer to the author's previous works on intelligent compaction. Several solutions have been developed, but detailed descriptions will be provided later due to business considerations.

As for automatic testing, technical difficulties have significantly diminished. The rapid advancement of electronic technology has allowed for the automatic collection, storage, and display of data. Testing challenges and processing techniques can now be better understood through direct experience.

Given the current state of modern science and technology, and with the status of intelligent compaction, it can be said that the foundational theoretical research related to perception technology has largely been completed, with only a few remaining areas for improvement. The perception technology used in intelligent compaction has now matured. Most problems that can be resolved using existing technology have already been addressed. At the same time, unresolved issues tend to be theoretical and do not affect the engineering application of this technology. These theoretical challenges may necessitate future scientific and technological breakthroughs.

(2) **In terms of "Analysis" and "Decision"**

Three key factors determine compaction quality: control of compaction degree, control of compaction stability, and control of compaction uniformity. Once a substantial amount of compaction data is collected, it is essential to analyze these factors using a specific analysis method (refer to Chap. 3) to determine if they comply with technical standards. The book provides detailed operational steps for this process. If a program is developed, the computer can perform automated analysis with ease.

The analysis follows established criteria. If the control criteria are met, the rolling process is deemed complete. However, if the requirements are not satisfied, rolling must continue. At this stage, a challenge arises regarding the compaction process: should the original process continue, or be modified? If a modification is necessary, what type of process should be adopted? This presents a decision-making dilemma in the compaction process. Currently, this aspect largely relies on experience; however, employing AI technology for supportive decision-making should be the focus of future development. This issue is a classification problem, and it is essential first to accumulate a significant amount of engineering experience, primarily represented as data in this text. The book outlines a technical approach that has not yet been fully implemented.

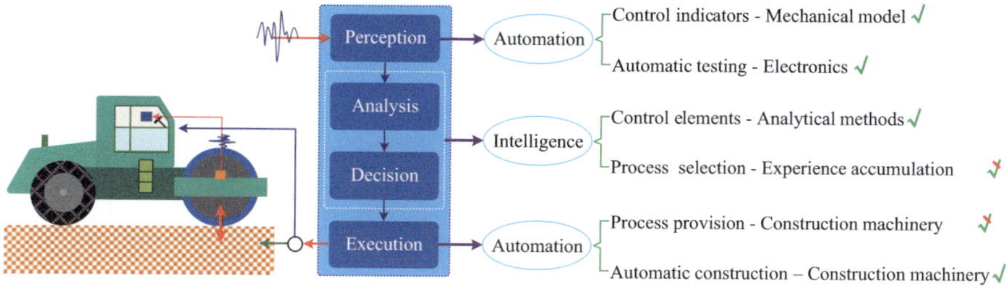

Fig. A.14 Current status of key technologies in intelligent compaction

(3) In terms of "Execution"

There are two key issues in execution. First, the roller needs to have a variety of compaction processes that can be automatically adjusted. Second, unmanned automatic rolling operations need to be implemented. The first requirement depends on advancements in engineering machinery, and some rollers already possess this capability. The second requirement has been achieved, like unmanned driving, although it is relatively straightforward and safe.

In summary, intelligent compaction has enhanced the theoretical foundations of control indicators and automated testing technology, enabling more accurate analysis and informed decision-making. The area currently lacking maturity, yet highly valuable, is selecting the appropriate compaction process. Intelligent decision-making regarding compaction processes shows great promise. While automatic rolling has been implemented, there are still relatively few compaction processes available. Figure A.14 illustrates the development status of key technologies in intelligent compaction.

32. What are the mechanical control indicators of intelligent compaction?

In traditional compaction quality control, two indicators are typically used: the physical indicator, density, and the mechanical indicator, modulus. These are two relatively independent indicators, and they do not necessarily have a strong correlation; otherwise, they cannot be considered independent (in elastic mechanics, density and modulus are among the most important physical parameters).

The most common use of density is to evaluate the stability of fine-grained soils that are sensitive to moisture content; the mechanical properties are most stable at the optimal moisture content (see Fig. 3.4). The method of obtaining density relies on in situ tests. Other methods are indirect methods. They are all statistical relationships established through comparative tests. They have certain limitations in practical applications, and there are also issues with accuracy.

The modulus describes the ability of the fill to resist elastic deformation—the larger the modulus of the fill, the stronger its ability to resist deformation. At the same time, the modulus is also a necessary parameter for mechanical analysis. The method of obtaining it relies on various mechanical tests, and the vibration compaction process is a dynamic test. For these reasons, modulus is used as a control indicator

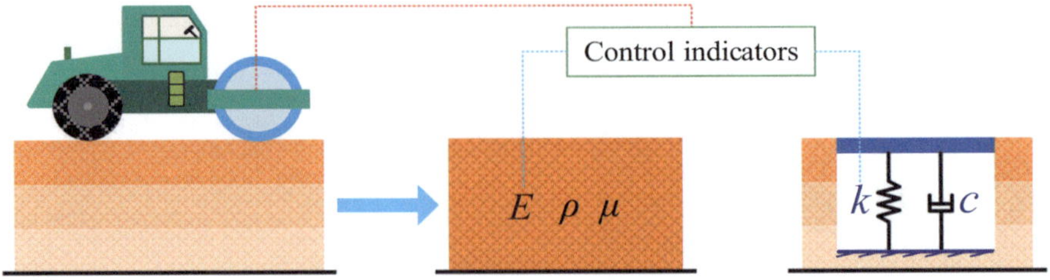

Fig. A.15 Two control indicators in intelligent compaction

in compaction quality control. One of the core issues of intelligent compaction is how to obtain modulus continuously and quickly. In addition to modulus, another mechanical index can be used to control compaction quality, namely the stiffness coefficient. Usually, the mechanical control index in intelligent compaction refers to modulus and stiffness coefficient. The following is a brief introduction to the differences and relationship between them.

Modulus and stiffness coefficient are parameters derived from various mechanical models. Modulus corresponds to the continuum model, and stiffness coefficient corresponds to the discrete body model, as shown in Fig. A.15. This book, along with references [1, 3], provides a detailed introduction to this issue.

There is no doubt that the filler is a continuum. Although there are some voids, they are ignored and modeled using elastic theory. Therefore, using modulus to control the compaction quality of the fill is an inevitable choice. However, the continuum model corresponds to a set of partial differential equations, that are difficult to solve and beyond the capabilities of general-purpose instrument manufacturers. In view of this, a discrete body model was proposed to simulate the compacted body (continuum) according to the principle of equivalence of force and deformation. Thus, two parameters, stiffness coefficient (spring coefficient) and damping, were found to simulate the elastic and viscous properties of the compacted body (see Chap. 2 and references [1, 3]). When controlling the compaction quality, the stiffness coefficient is selected as the control index (ignoring damping), which is easier than solving for the modulus.

From the perspective of compaction quality control, there is little difference between using modulus or stiffness coefficient as the control index, as both can achieve the purpose of controlling compaction quality. However, from the perspective of intelligent construction, considering the need to build a digital model of the compacted body and perform performance evaluation, the modulus is the best choice. Simultaneously, the stiffness coefficient is not considered an effective parameter and is rarely utilized. In addition, the stiffness coefficient is significantly influenced by the test equipment (such as roller parameters). Conversely, the modulus is an objective measure that remains unaffected by the testing apparatus (for ideal elastic bodies).

In addition to the stiffness coefficient, there is another similar control index, which is the foundation coefficient (K_{30}, where "30" represents the diameter of the bearing plate used for the test, which is 30 cm), which is the Winkler foundation reaction modulus, obtained by the plate load test method, and is a type of "resistance test

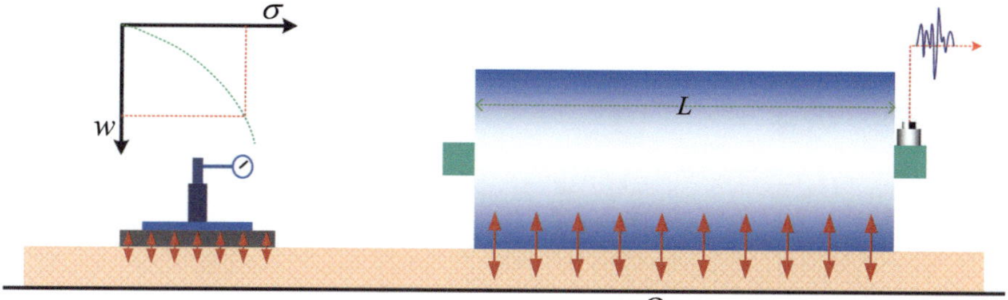

$$\sigma = \frac{Q}{S} \quad (\mathrm{KN/m^2}) \qquad q = \frac{Q}{L} \quad (\mathrm{KN/m})$$

$$K_{30} = \frac{\sigma}{w} = \frac{\sigma}{1.25} \quad (\mathrm{MPa/m}) \qquad K_{\mathrm{R}} = \frac{q}{w} \quad (\mathrm{KN/m^2})$$

Fig. A.16 Definition of foundation coefficient and stiffness coefficient

method". K_{30} represents the stress σ required to produce a deformation of $w =$ 1.25 mm, which is a method for evaluating the ability of the fill to resist deformation.

In addition, the resistance coefficient is also a control index, with a meaning similar to that of the stiffness coefficient (the two are essentially the same concept, differing only in name). To facilitate engineering applications, we can redefine the stiffness coefficient (also known as a resistance coefficient) for the vibratory roller, thereby minimizing the influence of roller parameters on the analysis. This coefficient is referred to as K_{Roller} (K_{R}). Figure A.16 explains the meaning of the foundation coefficient and stiffness coefficient.

The primary limitation of the K_{30} index is that it only measures resistance to deformation of up to 1.25 mm. Still, many fill materials do not reach a stable state when deformed by 1.25 mm (referring to non-elastic bodies, which are inherently non-elastic). Therefore, this index has certain limitations. The redefined stiffness coefficient K_{R} is a good index for controlling compaction quality, and it has a certain theoretical connection to the modulus. This index differs from the spring coefficient (the old stiffness coefficient).

It should be noted that, although intelligent compaction utilizes a robust mechanical control index, the detailed calculations and test methods are not disclosed, which limits the development of intelligent compaction to some extent. This situation needs to be changed. ISIC will be committed to achieving this goal (the author's R&D team is developing an economical mechanical index calculation module suitable for all vibratory rollers, which can be implemented in ordinary compaction monitoring systems to help relevant manufacturers improve their technical level. Please pay attention to the information on the ISIC public welfare expansion platform www.ISICSHOW.com, which will be launched soon.

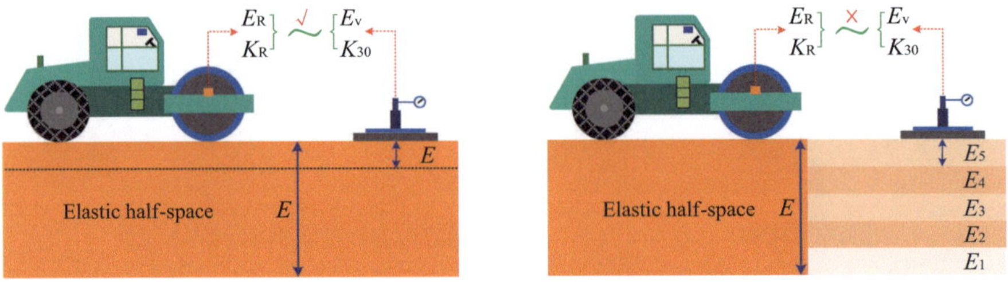

Fig. A.17 Comparison of the correlation of the two test results

33. Must the results of intelligent compaction tests be verified with those of traditional tests?

Currently, the technical standards for intelligent compaction (or continuous compaction control) in various countries all stipulate technical requirements for correlation verification, which involves correlating the results of intelligent compaction tests with those of traditional point tests, with a correlation coefficient of at least 0.70. The background for this requirement dates back decades, when technology was still at an elementary level (based on empirical control indicators). The primary concern was that empirical control indicators often failed to accurately reflect compaction quality, leading to cautious adoption of new technologies.

If advanced intelligent compaction technology based on mechanical control indicators (referring to modulus and stiffness coefficient) is used, correlation verification tests can be completely omitted. This is because the physical meaning of modulus and stiffness coefficient is very clear, unlike empirical control indicators. Of course, if you are willing to conduct correlation comparisons, it is also possible. Still, it should be noted that the correlation coefficient is sometimes very small (poor correlation), and the reasons are as follows.

As mentioned above (Sect. 5.6), the modulus (or stiffness coefficient) of the fill, obtained from the vibration response of the roller, is based on the elastic half-space theory and represents a comprehensive value within the test depth range. This test depth encompasses the thickness of several fill layers. However, the results of the plate load test only reflect the current fill layer. In other words, the test results of the roller include the test results of several fill layers (referring to the plate load test results), and the correlation between the two is closely related to whether the filler is uniform.

If the fillers in each fill layer are the same, the two test results exhibit a good correlation at this time, as confirmed by numerous engineering field projects. If the fillers of each fill layer are different (perhaps there is an uneven distribution of the fillers of the same fill layer in the horizontal direction), then the two test results may not have a good correlation (how to solve this problem, please refer to Sect. 5.6). Figure A.17 shows the two situations.

Additionally, it is worth reiterating that the test result obtained based on the vibration response of the vibratory roller is a mechanical quantity and cannot be used to detect or predict density. In principle, density can only be obtained through in-situ tests (such as the sand cone method and drilling method). Other methods are indirect and require calibration, and their accuracy is problematic. The idea of using mechanical tests (such as plate load tests and vibration compaction tests) to determine density remains a fantasy, at least at the current level of modern technology.

34. **How to use intelligent compaction technology in asphalt pavement construction?**

As a basis of the four basic needs of human survival, "clothing, food, shelter, and transportation", transportation infrastructure will gradually shrink as it reaches a certain scale. In contrast, maintenance and repair work will be vigorously developed. Asphalt pavement, the primary form of road structure, is likely to face numerous maintenance issues in the future. Therefore, the construction of asphalt pavement will increase, with paving and rolling being among the key tasks. Intelligent compaction (broadly speaking, including quality control during paving) will also be well applied in asphalt pavement construction.

(1) **Compaction control during paving**

During the paving process of the asphalt mixture, the first thing to control is the thickness and smoothness of the paving. The leveling system currently configured on the paver (including the laser ranging system and high-precision satellite positioning system) can automatically control the paving thickness and smoothness. Another thing that needs to be controlled is the segregation of the mixture (a widespread problem affected by factors such as material design, mixing, transportation, loading and unloading, and paving, and cannot be fundamentally solved at present). Here we mainly discuss the compaction control problem because improving the initial compaction degree will greatly help the subsequent rolling work (since the rolling of asphalt mixture is greatly affected by temperature, the higher the initial compaction degree after paving, the better).

While the paver is paving, the screed plate may also be vibrating to achieve preliminary compaction. However, this vibration compaction differs somewhat from that of the roller. It is a vibration in the form of a plate. Although its vibration response can be easily perceived, analyzing the vibration of the screed plate is challenging (the key is extracting the vibration characteristics), and there is also a modeling problem which may pose another difficulty.

According to our research results (2002), when the vibration intensity of the screed plate is high, its compaction degree is not necessarily high (i.e., the density does not necessarily increase). When the vibration of the screed plate exhibits certain characteristics but the vibration intensity is not very large (as reflected in both the time domain and the frequency domain), a higher initial compaction degree can be obtained. Although this vibration characteristic is difficult to analyze in theory, it can be handed over to AI for processing (AI is good at handling problems that are difficult to establish mathematical models). Figure A.18 illustrates the basic concept of controlling the vibration of the screed plate.

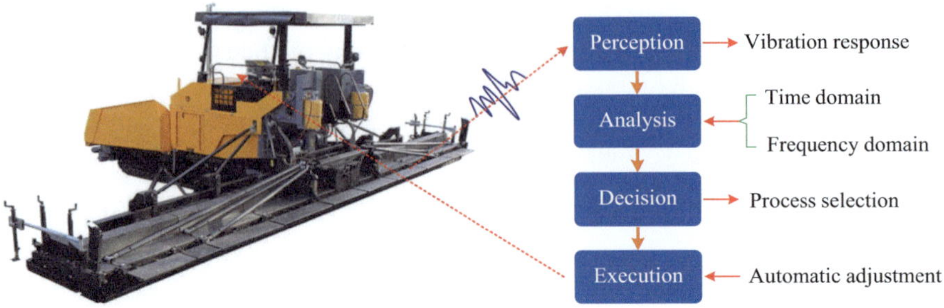

Fig. A.18 Intelligent compaction control concept of paver

(2) **Compaction control in rolling**

Regarding the compaction control of asphalt mixtures, the rolling pass control method is currently the most used. This is a traditional empirical method with extremely harsh conditions for use (see Fig. A.2). It does not belong to the category of intelligent compaction, nor is it intelligent. It is imperative to introduce advanced intelligent compaction technology in asphalt pavement construction. Here is a brief introduction to the general concept of the application.

Intelligent compaction originated with aggregate and soil fillers, and it presents two problems when applied to asphalt pavements. First, when a double-steel-drum vibratory roller is used for rolling operations, if the two steel drums vibrate at the same time, there will be mutual influence between them, and the established model is relatively complex and difficult to solve; second, since the change in temperature has a great influence on the compaction quality, it is necessary to consider the relationship between temperature T and the control index E_R or K_R (strictly speaking, it should be considered according to the high-temperature viscoelastic model, but it is relatively complicated).

For the first question, according to actual observation, at present, double-steel-drum rollers usually use one steel drum to vibrate during construction (if two steel drums vibrate at the same time, the phase between them needs to meet certain conditions, otherwise the vibration effect will be offset. This problem also occurs when multiple vibrating rollers are rolling in the same field. This lays the foundation for the mechanical model of intelligent compaction. In principle, the current intelligent compaction technology can be used (of course, due to the difference in roller vibration parameters, a control system needs to be developed separately). For the second question, practice has shown that at high-temperatures, the temperature's influence on the rolling effect is not significant. Still, there is a critical temperature $[T]$, which requires rolling operations to be performed under conditions above this temperature (rolling operations must be completed at a temperature above 80 °C, which is a critical temperature). Therefore, under high temperature conditions, it should be feasible to apply existing intelligent compaction technology.

In actual rolling, affected by several factors, the critical temperature $[T]$ is not necessarily 80 °C. It needs to be determined based on actual observation results;

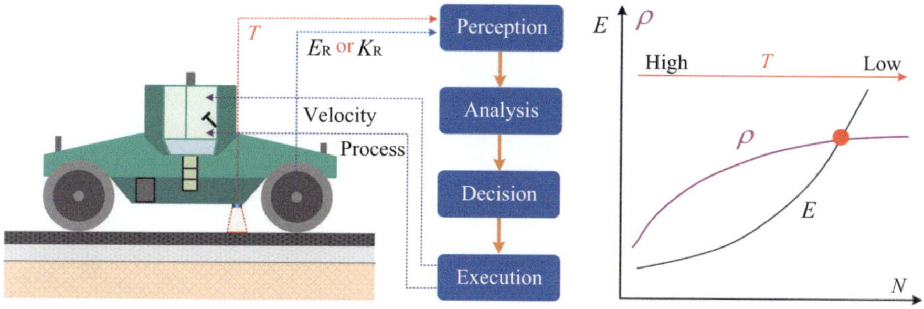

Fig. A.19 Intelligent compaction control idea for asphalt pavement

using AI methods may be a good idea (see the introduction volume of the series). Figure A.19 illustrates the concept of applying intelligent compaction control in asphalt pavement, where N represents the number of rolling passes.

In Fig. A.19, the intelligent compaction control system must simultaneously sense temperature and vibration signals. Specifically, the vibration signal from the roller drum needs to be converted into E_R or K_R. In contrast, the surface temperature signal T of the filling layer must be converted into an internal temperature signal. Before the temperature of the asphalt mixture drops to [T], intelligent compaction technology can be applied for control. At this time, the density ρ and modulus E will increase with the increase in the number of rolling passes N. When the temperature is below [T], the density will stop increasing, and the modulus will continue to rise. However, the compaction effect may become ineffective, or the asphalt mixture may not undergo further compaction. Therefore, in the construction of asphalt pavement, there is a problem of rolling operation time. It is necessary to adjust the rolling progress in a timely manner in response to temperature changes and the critical temperature, adjusting the rolling speed accordingly (v). Additionally, there is the issue of adjusting the compaction process.

It should also be noted that the curves of density ρ and modulus E versus rolling passes N (or temperature T) given in Fig. A.19 are only schematic diagrams. When applying intelligent compaction, it is necessary to measure according to actual conditions and determine the specific critical temperature.

(3) **Discussion on density measurement**

During asphalt pavement construction, density is a crucial topic, involving the issue of void ratio, which is a distinct characteristic of geotechnical materials. The technology for rapid density measurement has been studied; however, to date, in asphalt pavement, except for the drilling core sampling method, and no completely reliable indirect method exist.

The most classic indirect method for density measurement is the nuclear density method. Due to radioactivity and other reasons, the non-nuclear density method later appeared. They are commonly used to assist process control in asphalt pavement construction. In addition, in recent years, another indirect method has emerged—the

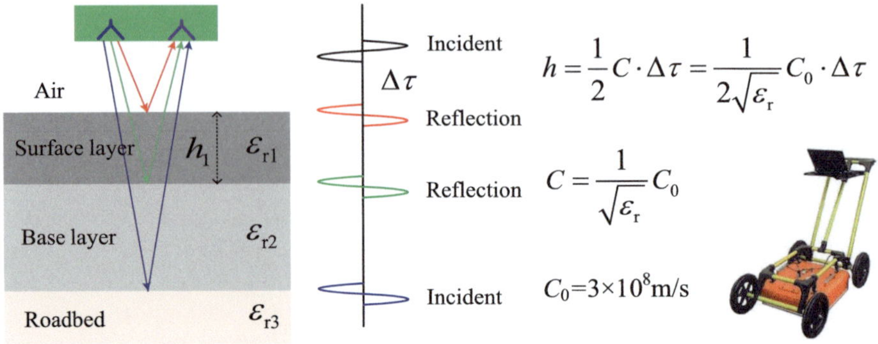

Fig. A.20 Basic principle of GPR detection of road structure

use of GPR technology to detect the density of asphalt pavement. In fact, this idea of using electromagnetic waves to detect density has existed for many years. At that time, it was used to detect the density of the roadbed, as well as the modulus and moisture content, among other properties. The detection mentioned here is referred to as "prediction," which may be a more accurate representation of the actual situation. The feasibility of this method is briefly analyzed below.

GPR stands for Ground Penetrating Radar, a method that uses high-frequency electromagnetic waves to detect the structure of the Earth's interior by reflection. Of course, it can also be used to detect internal defects in objects (or media), but the selected electromagnetic wave emission frequency is different. The dynamics part of the series has a detailed introduction, so I will not repeat it here.

In road engineering, GPR is used to detect the thickness of the road surface. It is based on the "distance measurement" principle of reflected waves. By emitting electromagnetic waves (incident waves) into the road surface, when encountering different media (different dielectric constants ε_r), the electromagnetic waves will be reflected at the interface. In this way, the distance from the emission point to each interface can be obtained according to the travel time $\Delta\tau$ and speed C of the electromagnetic wave, and the thickness of each layer of the road surface can be obtained (its basic principle is the same as the mechanical wave method), as shown in Fig. A.20.

Whether it is GPR or other reflection wave methods, the essence is to utilize the basic kinematic formula: distance = speed × time. The information received by the detection instrument is the reflected signal, from which the travel time $\Delta\tau$ of the electromagnetic wave can be determined. If the propagation speed C of the electromagnetic wave in the medium is known, the distance (thickness) can be calculated; if the thickness of the medium is known, C can be calculated.

Note that all applications of GPR are extended from here, among which the propagation velocity C is a key parameter, and its size is related to the dielectric constant ε_r. ε_r is a crucial parameter that characterizes the electromagnetic properties

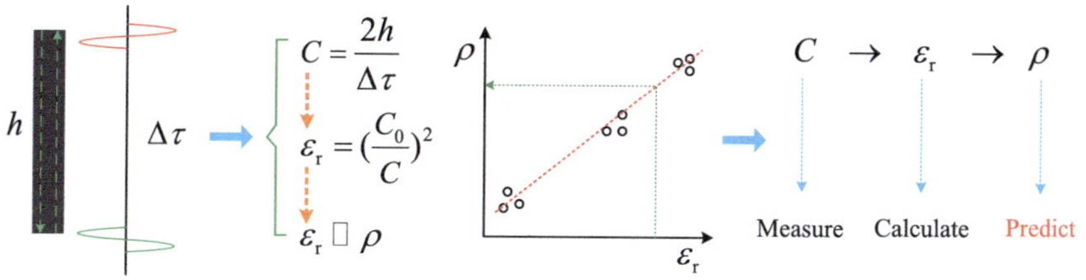

Fig. A.21 Basic principle of GPR prediction of density

of a medium. Its size determines the speed of electromagnetic waves propagating in the medium and is a key parameter for determining thickness and defect locations.

Due to the complexity of the asphalt mixture, its dielectric constant ε_r is variable, and there are many measurement methods (such as electrical measurement methods, electromagnetic wave measurement methods, etc.). Furthermore, empirical evidence indicates that the relative permittivity (ε_r) is also correlated with the density (ρ) of the medium. This point underlies the method of using GPR to predict density has emerged. This concept is like other indirect methods for predicting density. They are all empirical methods based on statistical significance.

To use GPR to predict the density of asphalt pavement, it is necessary first to establish a statistical relationship between the dielectric constant and the density of the asphalt mixture ($\varepsilon_r \sim \rho$). Specifically, core sampling is used, and GPR is employed indoors to measure the travel time $\Delta\tau$ of electromagnetic waves in the sample (height h). From this, the velocity C can be calculated, and subsequently, ε_r can be determined using the formula. If the density ρ of the sample is obtained, the statistical relationship between ε_r and ρ can be established, as shown in Fig. A.21.

Figure A.21 clearly shows the basic principle and process of using GPR to predict density. After the statistical relationship (statistical equation) between ε_r and ρ is established, ε_r can be calculated based on the measured wave velocity C on site, and then ρ can be predicted using the statistical equation

It is believed that if a strong statistical relationship is established, the density can be predicted with high accuracy. In fact, this is a very one-sided understanding! Although a statistical relationship with strong correlation can be established in the laboratory, the laboratory and the field are not the same. It is still necessary to be cautious when applying the results in the laboratory to the field (the boundary conditions are completely different). In actual on-site testing, many problems caused by the complexity of the dielectric constant ε_r distribution are difficult to solve.

For example, in the horizontal direction, the distribution of the dielectric constant is the same as that of the modulus and density. Each point on the filling layer may be different. The statistical relationship established from a small number of samples cannot fully reflect the on-site situation. In the vertical direction, there is a small possibility that the dielectric constant of each layer in the asphalt mixture is slightly different, which will lead to errors in calculating the wave velocity C. Consequently, the predicted density will also be incorrect, as shown in Fig. A.22.

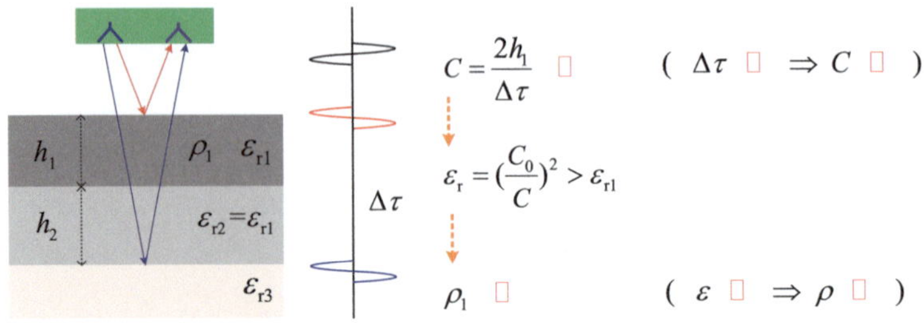

Fig. A.22 The Density prediction value is too large due to a small difference in dielectric constant

In Fig. A.22, if the dielectric constants of the upper layer (current filling layer) and the middle layer (underlying layer) are very close, that is, $\varepsilon_{r1} \approx \varepsilon_{r2}$, then the electromagnetic wave may not be reflected at its interface (the size of the dielectric constant determines the reflection strength of the electromagnetic wave. The greater the difference in dielectric constant, the stronger the reflection), but will be reflected at the next layer, which will cause the travel time $\Delta\tau$ of the electromagnetic wave to increase. Suppose the wave velocity C is still calculated according to the thickness of the upper layer (h_1). In that case, C will be artificially reduced, thereby increasing the calculated dielectric constant and the predicted density.

The above phenomenon also exists in rigid pavement. For example, the modulus and density of concrete and semi-rigid base differ. Still, their dielectric constants are remarkably close, making it impossible to distinguish between the two layers using GPR. Physical parameters (density and modulus) and dielectric parameters are not the same thing at all, and there is no absolute correspondence between them.

In summary, the question of whether there is a corresponding relationship between the dielectric constant and density should be addressed in physics. If we only start from experiments and use the dielectric constant to infer density, there is still a long way to go before it can be truly applied, and it still requires substantial data. In terms of density prediction, GPR will be similar to the nuclear density method (some methods have been developed using AI to predict density, etc.). It serves as an auxiliary means and is unlikely to become an accepted method for assessing compaction quality.

35. **What is intelligent construction machinery?**

The scope of construction machinery is very wide, encompassing a general term for various types of machinery used in engineering construction. It is generally divided into lifting machinery, transportation machinery, earthwork machinery, stone mining and processing machinery, road machinery, railway line machinery, concrete machinery, tunnel and bridge machinery, and other related machinery, such as excavators, loaders, graders, pavers, rollers, cranes, forklifts, concrete pumps, concrete mixers, concrete pump trucks, and special machinery. For transportation infrastructure, we primarily focus on road construction machinery, used to build roads and railways.

The interaction between machinery and medium determines the type of construction machinery. Road construction machinery is primarily categorized into three main types: earth and stone, asphalt mixture, and concrete construction machinery. For earth and stone, it includes machinery for stone mining, crushing and processing, transportation, paving, and rolling (including compaction); for asphalt mixtures, it includes machinery for aggregate manufacturing, mixing stations, paving, and rolling; for concrete, it includes mixing and paving machinery; in addition, there is road maintenance (watering, sweeping, snow removal, line marking) equipment, railway track laying equipment, etc. If these machines become intelligent, they should achieve intelligence in performance and function, which is also the need for intelligent construction.

① Intelligence of construction machinery performance. The intelligence of construction machinery performance falls under the category of intelligent manufacturing, primarily manifested as the ability to autonomously perceive the operating status of the machinery, master diagnostic skills through learning, and diagnose and maintain (or provide feedback information) any faults that occur. The application of electronic information technology to construction machinery can simplify the operator's operating procedures and improve the machine's technical performance. Using a controller, the machine's operating mode can be automatically selected. Additionally, sensing devices and microcomputers can autonomously perceive and detect mechanical faults, enabling autonomous maintenance and automatically detecting the quality of the machine's operation.
② Intelligence of construction machinery function. The function of construction machinery refers to its use, which is the ability that is manifested when the machinery interacts with the medium. Users of construction machinery are more concerned about the intelligence of its functions, which is the part most closely related to construction quality and efficiency. It is mainly manifested in the autonomous perception of the interaction information between the machinery and the medium (such as fill material, soil body), mastering construction skills through learning, making intelligent diagnosis of the object of action (medium), and automatically controlling the machinery to make the correct response according to the changes in the attribute parameters of the medium.

From the perspective of artificial intelligence, intelligent construction machinery is equivalent to construction robots. Various types of sensors are the sensory organs of construction machinery, responsible for obtaining certain attributes of the work object; the controller (microcomputer) is the brain and central nervous system of the construction machinery, responsible for receiving various information input by the sensors, performing intelligent calculations and analysis, deciding on action plans, and issuing execution instructions based on this; the execution mechanism includes electronic mechanical equipment (generally containing various motors), which can automatically perform corresponding operations according to instructions.

Intelligent construction machinery should incorporate humanization and spirituality, featuring sensory organs (sensors), a thinking brain (controller), and a limb skeleton (actuator), to enable automatic monitoring and intelligent control. Currently, many difficulties in the intelligentization of construction machinery are not entirely due to mechanical performance issues, but also to the correct perception of engineering information and the issuance of accurate control instructions. Technological breakthroughs in this area require in-depth research on the principles of interaction between machinery and media (operating principles or working principles). Intelligent compaction machinery is a good example.

The technological progress of construction machinery is inseparable from advances in its operating principles. For example, traditional operating principles include cutting, crushing, backfilling, and rolling of earth and stone, as well as screening, mixing, and heating of geotechnical materials. Additionally, these principles encompass spreading, leveling, and rolling of mixtures. These operating principles were originally based on bionics and Newtonian mechanics. With advances in science and technology, the operating principles of construction machinery have also undergone significant innovation. The application of mechanical vibration, pulse, jet, photoelectric effect, and thermo-physics has created conditions for the revolution of construction machinery, saved resources, improved engineering quality, increased production efficiency, work reliability, driving comfort, and automation level, and laid the foundation for the development of intelligent construction machinery.

There are many types of construction machinery, and the intelligentization of their performance shares many commonalities, which facilitates modular development. However, their functions are generally different, and there are several types of construction machinery with the same function, which is determined by their operating principles. Therefore, for different types of construction machinery, studying their interaction principles with the medium and then proposing functional requirements can promote the innovation of construction machinery products, drive technological progress in mechanical performance, and adapt to various functional requirements.

Taking the road roller as an example, looking at the development history of compaction machinery (see Fig. 1.6), each change in compaction equipment is closely related to its working principle (compaction technology and compaction method). The breakthroughs in the performance and functionality of compaction machinery all stem from advancements in compaction technology, which serves as the foundation for creating new compaction machinery. The development of intelligent road rollers will also follow this technical route.

In summary, technological progress in construction machinery and functional requirements are inseparable, and the expansion and improvement of functions often drive the innovation in construction machinery performance and types. Research on intelligent construction machinery must be a collaboration across several professions, encompassing a wide range of systems engineering, and there is still a long way to go.

36. **What is engineering automation?**

Engineering automation refers to the automatic operation of construction machinery according to predetermined goals without human direct intervention, aiming to replace human physical labor with machines (humans only indirectly supervise the

Fig. A.23 Requirements for automated operations in the spreading and rolling process

operation of machines). To achieve this desire, it is necessary to use computers, communication, and control technologies, etc., to design a set of control devices for the equipment (controlled object)—configure observers to measure (perception) output information automatically; configure controllers to process deviation information, analyze and make decisions, output control instructions; configure actuators to adjust the input of the equipment automatically. See the control technology volume of the series for details.

Engineering automation falls under the "object-object" type of control, which evolved from the "human-object" type (a driver operating construction machinery), in which the machine (object) replaces manual operation and becomes the primary means of control for construction machinery. In road and railway engineering, engineering automation primarily refers to the automation of construction machinery during the construction process of fill structures, encompassing the production stage of materials and the spreading and rolling stages.

In the production (preparation) stage of materials, automation has basically been achieved, including the production process of aggregates in road engineering, the mixing process of asphalt mixtures (feeding, mixing, discharging, etc.), and the production process of track slabs and rails (they are all prefabricated parts) in railway engineering etc.

In the spreading and rolling stage of materials (filler) (roadbed and pavement construction), the manual operation of mechanized construction is still mainly done. To achieve automated operations, you must meet the technical requirements outlined in Fig. A3.

Currently, the most popular and rapidly developing engineering automation projects include the automation of asphalt pavement laying and rolling, also known as unmanned construction (unmanned driving engineering machinery is a type of construction robot—a mechanical device that automatically completes construction tasks). This type of automated construction has two characteristics. First, it uses high-precision satellite positioning technology as its foundation, and its control principle of the walking trajectory is the same as that of unmanned driving cars (but the application scenario is simpler). Second, the control indicators are simple (to enable quantitative control), such as the walking trajectory of construction machinery, the thickness of the fill, the number of rolling times, etc., among which the walking trajectory is programmable control. The fill thickness and the number of rolling times are constant value controls. These two characteristics determine that the mathematical model of this type of automatic control system is relatively simple, and the corresponding control rules are also easy to formulate and implement.

For automatic rolling, if the roller cannot automatically recognize compaction quality and adjust the compaction effect during the rolling process, then the significance of automated construction is limited, and it becomes empirical construction that controls the rolling trajectory and the number of rolling times, eliminating human participation. The technology that enables the roller to recognize the performance of the fill automatically is intelligent compaction, an indispensable technology for automating fill construction.

37. **How to improve the technical level of automated construction?**

As mentioned above, automated construction (also known as construction automation) enables construction machinery to complete construction work without human intervention, utilizing automatic control (see the control technology volume). Can the technical level of automated construction be improved in this series? The construction machinery here is actually "construction robots", including rollers, pavers, graders, etc. In the field of highways, this technology first appeared in the construction of asphalt pavement (China, 2020), enabling automatic paving and rolling operations of pavers and rollers without driver intervention. Later, similar automated construction scenes appeared in other countries.

The current level of automated construction is not high. It primarily introduces various technologies related to autonomous driving in automobiles. The machine (control system) initially replaces the various operations of the driver and realizes unmanned driving. Since the operation is in a closed site, personal safety is less of a concern (this is much safer than driving an automatic vehicle). Still, it does not significantly enhance the project's quality. For example, for the rolling process, the original roller (just transformed into unmanned driving) and the original compaction process are still used. Construction progress and quality are controlled by recording the number of rolling times, which is still far from people's expectations (for improving work efficiency and project quality). Nevertheless, this development trend is still worthy of recognition. At least the first step has been taken on the road to automated construction.

According to the technical requirements of intelligent construction (see the introductory volume of the series), ideal automated construction should possess basic characteristics such as automatic perception, intelligent analysis and decision-making, and automatic execution, with the primary goals of enhancing project quality, improving working conditions, and increasing work efficiency. Given that automated construction is still in its early stages of development, it is necessary to categorize it to advance this technology.

(1) **Technical classification**

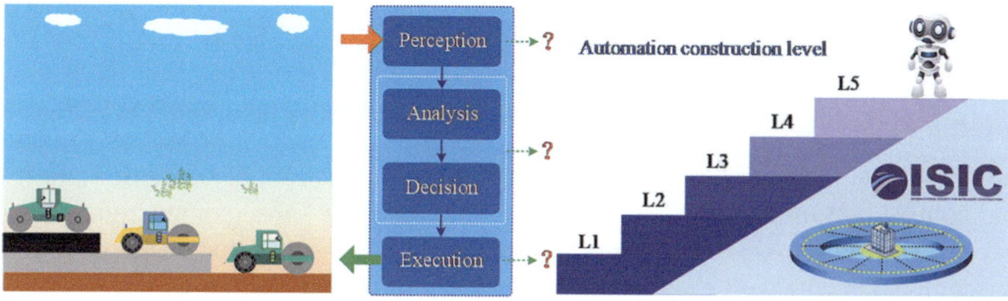

Fig. A.24 Classification of construction automation levels

In the industrial world, although the automation of the entire product production process has been realized for a long time, it encompasses material preparation, loading, manufacturing, inspection, and other stages. However, upon reviewing its development history, it was found that only some links were automated in the early days, which was at a low level of automation. Later, as the technology advanced, the level of automation continued to improve.

Inspired by this, the engineering industry should also gradually increase automation in the construction process. Its development speed is determined by the pace of related technologies, which leads to the issue of the development level of automation, namely, the issue of classification.

To classify the level of automation in construction, it is necessary first to determine the basis for this classification. Under the premise that engineering machinery has achieved automatic driving (i.e., automatically sensing geographic location and surrounding information, adjusting driving direction and speed, etc.), the level division is primarily based on the following two points. First, the degree of improvement in engineering quality; second, the degree of improvement in work efficiency. In actual operation, if the degree of automation and intelligence in "perception, analysis, decision-making, and execution" is grasped, it can be divided.

Based on the above principles, we mainly focus on the compaction stage, and according to the specific content of "perception, analysis, decision-making, and execution", the technical level of construction automation is divided into five levels, as shown in Fig. A.24. The following is a brief explanation.

Level L1—Except for the uncrewed operation of the roller (in many cases, remote control assistance by off-site personnel is required), there is not much difference from the traditional construction method in other aspects; that is, rolling is still carried out according to the traditional process (with relatively few adjustable process parameters), using level 1 intelligent compaction technology. In addition to automatically sensing the information required for automatic driving, the roller also automatically senses the geographical location of the rolling surface, based on which the number of rolling passes is counted and compared with the specified numbers. When the specified value is reached, the rolling is automatically stopped. At this level, there is only simple automatic perception and execution, and no intelligent analysis or decision-making is required. In terms of work efficiency, the machine only replaces the driver's work, and the project quality is essentially the same as that of traditional construction.

Level L2—The roller is the same as the L1 level, and the intelligent compaction is level 3 technology (the control indicator is the modulus E_R or stiffness coefficient K_R), primarily used to control the degree of compaction, compaction stability, and compaction uniformity. The roller can automatically sense compaction quality information, analyze control factors, and participate in the decision-making process for compaction, as well as perform corresponding operations. At this level, work efficiency remains the same in L1, but compaction quality improves.

Level L3—The roller features a variety of compaction processes that can be automatically adjusted and incorporates intelligent compaction technology of level 3 or above, primarily to control compaction degree, stability, and uniformity. The roller can automatically sense compaction quality information, analyze control factors, determine the compaction process, and perform corresponding operations. At this level, the work efficiency is higher than that of level L2, and the compaction quality is also higher than that of level L2.

Level L4—Building on Level L3, the roller is further integrated with advanced intelligent compaction technology, that features self-learning capabilities. It can intelligently diagnose various problems that occur during rolling and provide intelligent solutions. Work efficiency and project quality are further improved.

Level L5—Automation and intelligence are realized in the entire process of construction, including material preparation, loading, paving, rolling, acceptance, and other links. The construction site is truly unmanned, and digital twins are realized through the interaction of various engineering machinery and testing equipment with the information platform. Human engineers can assist in monitoring, coordinating, and directing the collaborative operations of various pieces of engineering machinery on the construction site through digital models on the information platform.

It should be noted that the division here is mainly based on the compaction stage. Because in the filling project, the compaction stage is the core, and its level of automation determines the level of automated construction. Additionally, the following is a brief description of the technical classification. We will provide a more detailed division and description under ISIC. The following introduces the technical approach to developing high-level automated construction.

(2) **Technical approach**

For intelligent construction, it is natural for us to develop high-level automated construction. From a technical point of view, integrating and fusing of multiple technologies is the key to improving the level of automation, and it is also the only way. In fact, in the currently popular L1 level, it is the integration of related technologies. For example, the unmanned driving part involves integrating related technologies from automotive autonomous driving. In contrast, the technology for automatic operation of engineering machinery comes from the mature technology of the manufacturing industry (engineering machinery itself is a product manufactured by the industrial sector). Therefore, in improving the level of automated construction, this technical

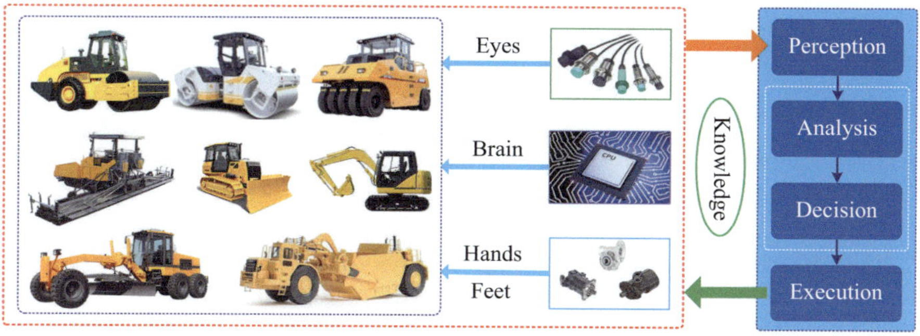

Fig. A.25 The basis for engineering machinery to achieve automation and intelligence

route can still be adopted to widely absorb mature technologies from related industries and apply them for our benefit. Of course, independent research and development are also necessary, but they require a significant workforce and material resources, as well as substantial time costs.

Engineering machinery was invented as a tool to replace human physical labor and is a key component in construction. In the early days of today's intelligent era, we require engineering machinery not only to "work automatically", but also to "work" and "work well", which corresponds to different levels of automation and intelligence.

Therefore, it is necessary to equip engineering machinery with "brains", "eyes", and "hands and feet", and then teach it relevant professional knowledge so that it can automatically "perceive, analyze, decide, and execute". This is the basis for engineering machinery to achieve automation and intelligence, as shown in Fig. A.25.

For engineering machinery, the "brain" is the computer, the "eyes" are the sensors, and the "hands and feet" are the motors. Given the current level of technological development, it is relatively straightforward to configure these hardware facilities for construction machinery. In other words, it is not difficult to enable construction machinery to work automatically. The difficulty lies in teaching them how to work effectively and do an excellent job. There are both professional issues and machine learning issues (mainly involved in L4 and above technologies), and different types of construction machinery involve specific content.

In filling projects, rollers are the key protagonists of construction, and compaction quality is paramount. Therefore, this type of project can initially implement construction automation around rollers and then gradually expand to other mechanical equipment, such as pavers, graders, bulldozers, and excavators. The current situation is developing in this manner.

Rollers' primary function is to roll along a specified route and stop rolling when the required conditions are met. At the L1 level, the roller is confined to a specific area, following the planned rolling route and the number of rolling passes. By automatically sensing the geographical location and surrounding condition, a simple automatic analysis and decision-making process is performed, and then the actuator is used to achieve automatic rolling. To improve compaction quality, the level of automation can be raised to Level 2 by introducing advanced intelligent compaction technology above Level 3. If you want to continue improving your level, it becomes difficult to rely on this traditional roller. It is necessary to introduce rollers capable of multiple compaction processes and automatically adjustment of process parameters, thereby leveraging the advantages of intelligent compaction more effectively. AI technology will also be introduced in large quantities.

It should be noted that if the compactability of the filler is very good (see Fig. A.3), the L3 level technology can generally be used to complete automatic rolling, thereby improving its engineering quality and work efficiency. For fillers with average or poor compaction (to be realistic, this type of filler does not meet the requirements of the specification, but is widely used in practice), the situation will become more complicated. AI intervention is required currently (it is necessary to be familiar with the corresponding algorithm and be able to apply it flexibly to the rolling work). Only by making the roller "smarter" (including the technology of intelligently identifying the compaction characteristics of the filler) can the rolling task be completed. For details, refer to the discussion in the relevant chapters that follow.

For projects such as roadbed and base layers that are mainly composed of aggregate and soil, the number of rollers required for rolling operations on site is relatively small, and there is essentially no need for multiple rollers to work together. For asphalt surface layers, multiple rollers are typically used simultaneously for rolling operations. Currently, it is necessary to consider the problem of coordinated operation, as well as between rollers and pavers, pavers and loading vehicles, and loading vehicles and mixing stations. They constitute a complex construction Internet of Things system, as shown in Fig. A.26. Its coordinated control problem is a challenging research direction.

It should also be noted that the two construction photos given in Fig. A.26 are both L1-level automated construction. Although automatic paving and rolling has been achieved, it is not particularly meaningful unless it is combined with advanced intelligent compaction technology. Because the roller needs to understand the rolling quality at any time, it must also learn to adjust compaction process parameters and automatically implement adaptive control. This is necessary for automated construction, and controlling the number of rolling passes is not enough.

(38) How to make construction machinery have adaptive control capabilities?

Automated construction has three meanings. First, the movement and operation of construction machinery are automated. Second, the interaction information between construction machinery and construction objects (such as fillers, foundations, slopes, etc.) is automatically sensed. Third, construction machinery can automatically adjust its operating mode in response to the changes in the construction object. The first two problems have been solved. The following describes how to solve the third problem, automatic adjustment of the operation mode of construction machinery, thereby enabling adaptive control capabilities (this is a problem that determines the level of automated construction).

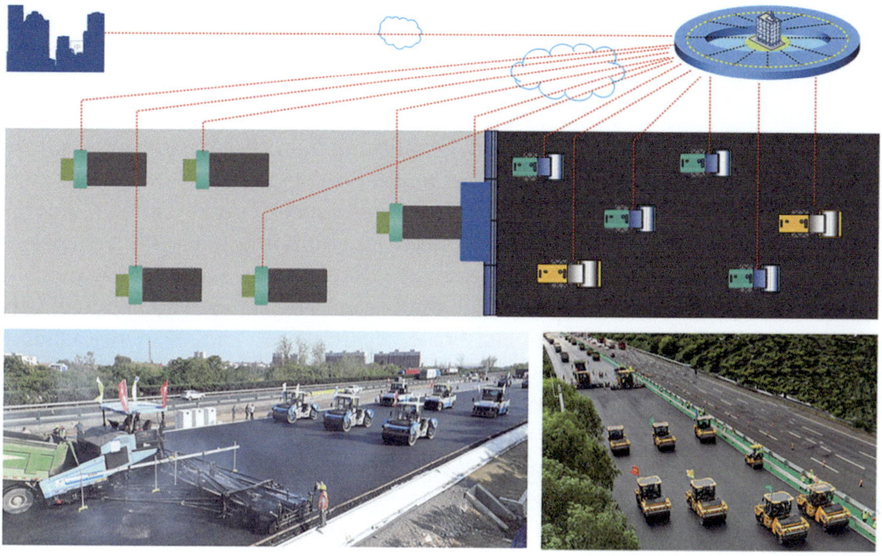

Fig. A.26 Construction internet of things system

We still use the roller as an example. Because the roller is not only the protagonist of filling construction but also a complex type of construction machinery, solving the adaptive control problem of the roller will also inspire solutions to the adaptive control problems of other construction machinery.

During the rolling process, the roller continuously compacts the filling material, and the parameters representing its physical and mechanical properties change constantly (reflecting changes in compaction state). The roller can automatically sense changes in the mechanical performance parameters (E_R or K_R). If this change is regular, then it will be quite simple to achieve automatic rolling control (a common automatic control problem). The traditional rolling control method is based on the idea that the test section determines changes in performance parameters of the filling body. Then the corresponding rolling passes are determined when the rolling is qualified.

Due to the complexity of the filler and the influence of various factors, it is difficult to determine the change law of the performance parameters of the filling body in advance (the rolling passes obtained through the test section has no universal significance), which means that the roller does not know how the performance parameters of the filling body (referring to E_R or K_R) change in advance, and it is impossible to formulate an exact control plan. The roller must continuously adjust the compaction process parameters in response to changes in E_R or K_R information during the rolling process to adapt to changes in the compaction state. This is adaptive control (see the control technology volume of the series for details).

Let the roller (construction machinery) automatically adjust the compaction process (action mode) based on changes in the filling body (construction object).

This has always been the goal pursued by intelligent rollers (intelligent construction machinery), and it is also a skill that L4 and L5 levels should possess. Since the change of fill performance cannot be represented by an exact mathematical model, the traditional automatic control method cannot be implemented. Adaptive control involves two problems. The first is which compaction process parameter to adjust-whether to adjust the exciting force or the frequency; the second is how to determine the adjustment range of the parameter. These two aspects are key to enabling the roller to achieve adaptive control capabilities. One solution is to use AI technology, which excels at solving complex problems without relying on mathematical models.

We know that in autonomous driving, the reason why the car can automatically adjust the steering drum, throttle, and brake according to changes in surrounding information is that the car (the control system in it) simulates the driver's adaptive control behavior. The driver's driving skills are acquired through learning and practice. Similarly, a roller equipped with an intelligent compaction control system can acquire the ability to adjust the compaction process through machine learning, thereby instilling it with corresponding professional knowledge and enabling adaptive control.

For rollers, there are two machine learning. One approach is to learn before doing—build a compaction knowledge base, select a suitable algorithm, let the roller learn the knowledge in the knowledge base first, and then work after acquiring the necessary skills. The second approach is to learn while doing it. This method is based on reinforcement learning, which incorporates on-site rolling as part of the learning process, allowing for learning while rolling, or extracting knowledge from the compaction knowledge base for virtual rolling learning.

These two machine learning methods are introduced in this book and the introductory volume of the series, so I will not repeat them here. What needs to be paid attention to is the compaction knowledge base, which serves as the foundation for determining whether the roller possesses operating skills. It is also one of the sources of "big data" in the filler engineering. This knowledge is inseparable from the integration of "materials-design-construction". When building a compaction knowledge base, it is essential to consider the following two key aspects.

First, the compaction knowledge base should include different types of fillers, gradations, moisture content, corresponding compaction processes (including roller types, corresponding compaction process parameters, that is, vibration mass M, excitation force P, vibration frequency f), compaction effects (compaction degree, compaction stability, compaction uniformity), fill layer thickness, and underlying layer information. If it is an asphalt mixture, in addition to material properties, compaction processes, and compaction effects, temperature information during rolling should also be added. In short, the more comprehensive and richer the content contained in the compact knowledge base, the more convenient it is to use.

Second, the traditional knowledge base is based on an expert system; however, the expert system cannot learn, which is also its fatal disadvantage, resulting in longer update times and higher maintenance costs. To address the shortcomings of the traditional method, it is necessary to integrate machine learning with the expert system to develop a hybrid knowledge base, as illustrated in Fig. A.27.

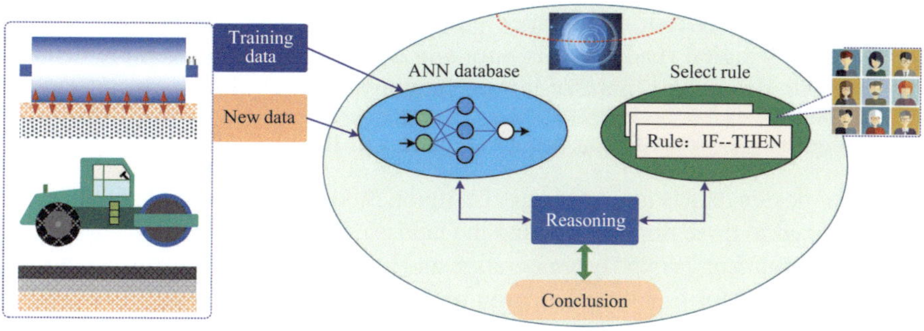

Fig. A.27 Hybrid compacted knowledge base

In Fig. A.27, part of the knowledge is represented by a trained artificial neural network (ANN). The knowledge is stored in the weights of the neuron connections, which is an implicit expression. An ANN model can represent a set of related knowledge (a set of rules) and can automatically obtain new weight distributions and update the knowledge base based on new training data.

For rules that are difficult to represent with an ANN, IF-THEN can still be used. For fuzzy problems, fuzzy-expert systems can be employed, while probabilistic-expert systems are suitable for non-deterministic problems. In addition, it is worth studying how to effectively utilize large language models (such as ChatGPT and DeepSeek).

Currently, almost all construction machinery can operate autonomously (automatically), but the quality of the operation is another matter. It is crucial to employ various machine learning methods to equip construction machinery with learning capabilities, particularly adaptive control. Its research value is exceedingly high. One key is to build a corresponding knowledge base so that construction machinery has sufficient learning materials.

39 **What is the relationship between intelligent compaction and virtual construction?**

The role of intelligent compaction in filling projects is well known. It also plays a key role in virtual construction. Some introductions have been made in Sects. 5.6 and 5.7 of this book. Here are some additional explanations.

The so-called virtual construction is to build a virtual structure (digital structure) in the computer (virtual world) and perform various operations on it, just like growing flowers and vegetables in video games, forming a one-to-one correspondence between the physical world and the virtual world (see the introduction volume of the series). In the intelligent construction of transportation infrastructure, virtual construction has significant potential and numerous applications. Taking road engineering as an example, first, understand the purpose of virtual construction, and then introduce the relationship between intelligent compaction and virtual construction.

In the virtual world of a computer, a digital road is built. Many things can be done with this digital road. Here is only a brief explanation. For a more detailed introduction, please refer to the series' perception and data volume.

(1) **Design stage—evaluating design solutions in the virtual world**

If detailed design data is imported into the digital model, the alignment and structure are visualized in three dimensions, and the landscape along the route (including the natural environment, primarily temperature and humidity) is superimposed, creating a virtual road that forms a virtual environment for dynamic simulation. For example, simulation work can be carried out in three aspects: alignment (related to vehicle driving), structure (related to the life cycle), and landscape (related to the driver's feelings).

Displaying the shape of the road in three dimensions in the virtual space is equivalent to three-dimensional design, which can change the traditional two-dimensional design method that separates the plane, elevation, and section, making the design results more intuitive and accurately expressing the geometric characteristics of the road. More importantly, driving experiments can be conducted on virtual roads (operated by humans or machines) to evaluate the advantages and disadvantages of the alignment from the perspective of vehicle kinematics and obtain parameters useful for alignment design.

Another important simulation task in the design stage is to evaluate the structure (including materials) of the proposed road, particularly to predict the its lifespan, ensuring it meets the design life requirements. Virtual simulation experiments can control time. Suppose the model for calculating the fatigue of road structures is accurate and has complete data on vehicles and the natural environment. In that case, it only takes a brief time to simulate the fatigue life of the road under driving loads for more than ten years or even decades, thereby understanding the durability of the road surface.

(2) **Construction stage—Synchronous growth of entity and virtual body**

Construction is the process of transforming the designer's ideas (design drawings) into real roads (products), marking the beginning of a new life for roads. The data in this stage is the richest, not only in terms of many types but also in massive quantities; however, not much construction data is preserved. Most of the data, especially data related to the construction process, disappears after the construction is completed. Suppose the data from the construction stage is stored in a digital model. In that case, it can not only trace history but also reproduce the construction process, making it convenient for data analysis, and greatly beneficial for the comprehensive use of data.

Suppose the digital model in the design stage is "paper talk" relying mainly on historical data. In that case, the digital model (virtual body) in the construction stage grows synchronously with the road (physical) construction process. With the help of the Internet of Things, various engineering data from the construction site can be received in real time. Here, the digital model is not only equivalent to a data lake, which can store diverse types of data, but also enables the visualization of different data stored in the corresponding parts of the virtual body. For example, the type and gradation of the base material are stored in the base position, and the performance data of the base structure are stored at various points along the base (the data at each point may differ), which is convenient for future data retrieval.

The road structure is formed by filling and rolling various materials. Since intelligent compaction technology enables continuous detection during the rolling process, the modulus at each point in the filling body can be obtained, thereby injects the performance data of the entity into the digital model, rather than simply virtualizing the road geometry.

Based on our practical experience, it is currently possible to simulate the filling and rolling process in the virtual world and transmit on-site construction data to the virtual road structure in real-time. The construction data includes modulus, material type, location information, and compaction process parameters for each point. These data are stored in the corresponding parts of the virtual body and correspond to the actual filling body. In this way, whether it is a physical or virtual body, its state is known, providing data support for the operation and maintenance of the road. At the same time, the information platform can analyze the degree of compaction, compaction stability, and compaction uniformity based on the compaction data transmitted, and the results can be fed back to the entity in real-time, forming a virtual-real interaction.

Once all the filling and rolling work is completed, the corresponding digital model is formed, along with the road structure. This marks the beginning of a new life on the road, and the initial state is known. When the road is handed over to the highway management department, the digital model should also be transferred to ensure that data continues to flow.

(3) **Operation stage—synchronous work of the entity and virtual body**

In the operation stage, from the perspective of road engineering, we are mainly concerned about the mechanical response and damage of the road structure (this is also the task of the maintenance stage); from the perspective of traffic engineering, we are mainly concerned about the driving problems of vehicles, among which intelligent traffic is the focus. These tasks can be carried out on the virtual road that has been handed over, including mechanical analysis of the road structure, damage identification, service-life prediction, and management and control of vehicles on the virtual road.

The operation stage is the growth stage of road life. The update of the digital model primarily consists of two categories: measured data from real roads and calculated data. In the virtual world, we can use various data analysis techniques to conduct a comprehensive analysis of this multi-source data and provide feedback to the physical road. These are also tasks to be completed during the maintenance stage.

(4) **Maintenance stage—make maintenance decisions in the virtual world**

The maintenance stage is closely tied to the operation stage. Since the virtual road incorporates measured data and calculated data from the real road, various data analysis and simulation tasks can be performed on this digital model, and virtual-real interaction can also be facilitated. For example, changes in the actual road state can be observed on the virtual road, and some predictions (such as performance degradation, damage, and life expectancy) can be made. Suppose the engineer wants to view a specific part of the road. In that case, they only need to equip themselves with professional glasses, and the digital model can provide the information they need. This augmented reality (AR)—superimposing information about the virtual road structure (such as defects) onto the actual road on site, so that the on-site engineer has the "perspective" function.

Additionally, depending on specific needs, machine learning models can be trained in the virtual world using data from the physical world, enabling intelligent analysis and decision-making within the virtual environment.

The primary task of the maintenance stage is to develop a maintenance plan based on various diagnostic data from the road (like determining a treatment plan based on human diagnostic information in medicine, as some have proposed referring to the road as a "road hospital"). There are many benefits to making maintenance decisions on virtual roads. For example, not only can the decision-making process in the physical world be moved to the virtual world, but the maintenance process can also be repeatedly simulated on the virtual road to find the best solution, which is simlar to the design stage. In fact, maintenance is a small-scale secondary construction that also requires design and the transfer of relevant data to the design stage.

The above briefly introduces some uses of virtual construction in the design, construction, operation, and maintenance stages. However, at present, this type of technology still has a significant limitation—it cannot simulate the performance of the structure virtually. It cannot conduct simulation experiments of the entire life-cycle process. The reason is insufficient supporting data.

For filling projects, continuous detection in intelligent compaction can provide performance data (modulus) at every point of the structure, offering unique advantages that other technologies do not have and may hold promise in solving this problem. Figure A.28 illustrates the relationship between intelligent compaction and virtual construction of filling projects, highlighting the importance of data in this connection.

According to system theory, in the physical world, all contents of the four stages of road engineering design, construction, operation, and maintenance can be integrated into a large system, including multiple elements (personnel, equipment, vehicles, materials, methods, environment, management, etc.), covering the entire life cycle of the road. Virtual construction involves moving this entity into the virtual world and performing various virtual operations. From the current situation, it is acceptable in terms of virtual appearance and scene (like the practice of video games), but it is not very useful in practice. It can only be used for demonstration (outsiders may prefer it), but it is not strong in other areas (such as performance and function simulation, etc.). The primary reason is the lack of support from substantial amount of actual data, and intelligent compaction can address the issue of obtaining fill performance data.

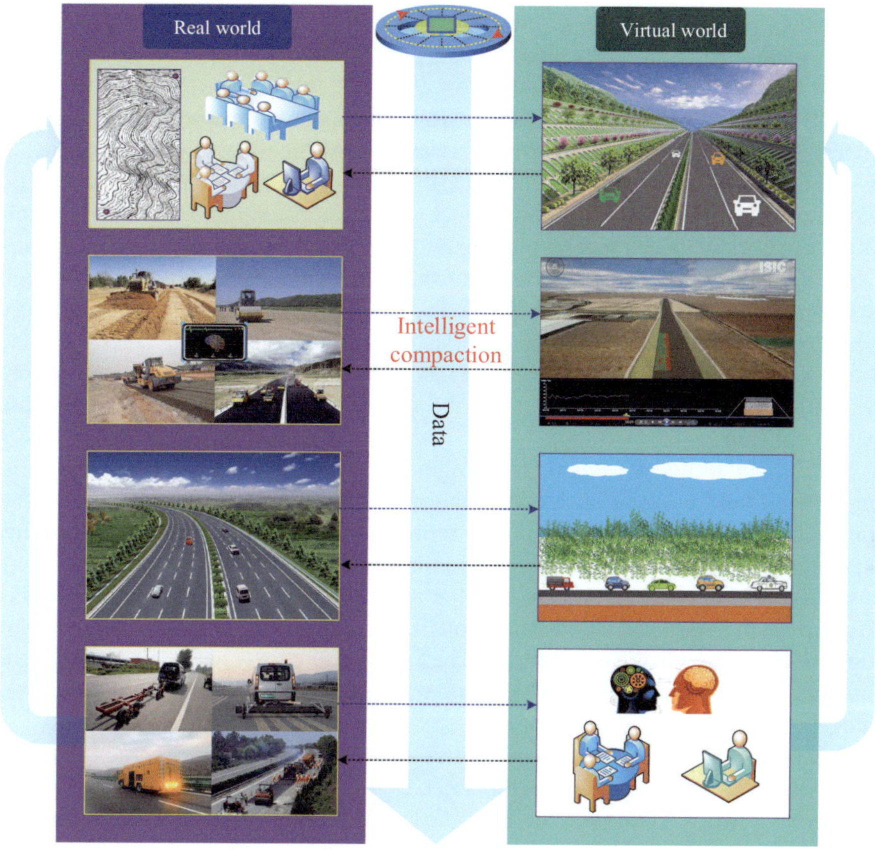

Fig. A.28 Relationship between intelligent compaction and virtual construction

40 What is the future development direction of intelligent compaction?

Complete intelligent compaction should include four basic processes: "perception, analysis, decision-making, and execution". This is different from continuous compaction control, and its development direction should also be discussed from these four aspects. As a pioneer in intelligent construction, intelligent compaction has initially achieved autonomous "perception, analysis, decision-making, and execution," and has begun to play a significant role in engineering construction. Figure A.14 shows the status of intelligent compaction. The part that has not been fully resolved is the future development direction, which is mainly reflected in "analysis, decision-making, and execution". The following is a brief explanation.

For intelligent compaction, the "execution" process is the "control" process, "decision-making" serves control, and "analysis" serves decision-making. Their common basis is "perception", that is, obtaining data. After understanding these relationships and combining them with the status (see Fig. A.14), it is not difficult to see that the key development direction of intelligent compaction in the future is how to make decisions on the compaction process, which can be divided into two specific research routes.

First, in theory, the strength and mode of interaction between the roller and the fill determine the compaction effect, which is based on the theory of dynamics. There are not only the dynamic problems of engineering machinery, but also the elastic–plastic dynamic problems of the filler. The essence of the compaction process is the combination of roller vibration mass M, the exciting force P, and the vibration frequency f. Its essence is the size and frequency of the compaction force. For materials such as crushed stone, soil, and asphalt mixture, the compaction process is relatively complicated, and the applied force needs to change according to the changes in the compaction process. However, quantitatively describing this change and determining the appropriate amount of force (i.e., the compaction process) requires in-depth theoretical studies.

Second, in terms of experience, engineers who have worked on long-term engineering field projects have accumulated extensive experience in adjusting the compaction process. However, with the retirement of older engineers, much of this experience has been lost, which is something we need to address. The main way to solve this kind of problem is to develop and introduce AI technology as soon as possible (such as developing an intelligent expert system) and turn the experience of engineers into knowledge and store it in the knowledge base. This forms the basis for applying adaptive control technology, and numerous specific details require further study.

Comparing the above two research directions, it may be more practical to transform traditional experience into AI models, as it is challenging to conduct in-depth theoretical research using existing scientific and technological means unless a breakthrough in science occurs. Therefore, the integration of intelligent compaction and AI will be the main research direction in the future. We will release relevant information on the upcoming ISIC expansion public welfare platform www.ISICSHOW.com, so stay tuned.

Appendix B
Classification of Intelligent Compaction Technologies

This appendix explains why IC technology classification is necessary, how to perform IC technology classification, and provides some example systems/solutions.

1. Background

Intelligent compaction (IC) is an equipment-based technology that improves the quality control of compaction. IC vibratory rollers are equipped with a high-precision global positioning system (GPS), infrared temperature sensors, an accelerometer-based measurement system, and an onboard color-coded display. IC is used to improve compaction control for various pavement materials, including granular and clayey soils, subbase materials, and asphalt materials. The accelerometer-based measurement system is a core IC technology that was invented in the early 80 s and is still evolving today.

Intelligent Compaction Measurement Value (ICMV) is a generic term for an accelerometer-based measurement system instrumented on vibratory rollers as a key component of intelligent compaction systems. ICMV is based on the acceleration signals that represent the rebound force from the compacted materials to the roller drums. ICMVs are expressed in various forms of metrics with different levels of correlation to the mechanical and physical properties of compacted material, including stiffness, modulus, and density.

ICMVs are expressed in various forms of metrics with different levels of correlation to the mechanical and physical properties of compacted material. The purpose of this document is to demystify ICMV by providing a comprehensive description of its mechanisms and various levels of solutions, serving as a roadmap for using ICMV in compaction monitoring, control, and acceptance.

To assist the academic and industry in developing their ICMVs, the US Federal Highway Administration (FHWA) Tech Brief (FHWA HIF-17-046), ICMV—A Road Map, provides the most comprehensive description and classification of the ICMV for five levels (FHWA, 2017). The primary goal is to dispel several myths about ICMV that may hinder industry adoption, thereby limiting ICs from widespread use despite over 50 years of development. However, the ICMV classification does not

G. Xu et al., *Pioneer of Intelligent Construction - Intelligent Compaction*,
https://doi.org/10.1007/978-981-95-4184-3

discourage low levels of ICMVs. Each ICMV level has its application and historical significance. The following explains the ICV mechanism, various ICMV solutions, and the factors that challenge the computation of ICMV.

2 How to Measure and Calculate ICMV

Comprehensive modeling of an IC system requires simulating the interacting system, which is composed of the vibratory roller and asphalt or foundation layers. The available literature on the simulation of IC systems consisted of approaches for modeling the roller components and equipment or simplified to enhanced analysis approaches considering the soil-drum (or asphalt-drum) interaction. The model-based analysis of IC systems, when appropriately calibrated against field data, can facilitate the implementation of high-level IC and, in turn, accurately predict pavement performance.

Aside from modeling, implementing IC technology poses a challenging issue that transportation agencies must address. To this end, IC specifications have been developed by many countries to ensure quality inspection, increase construction efficiency, improve pavement performance, and enhance work safety.

The demand for the real-time display of ICMV within 0.1 s (typically when a roller moves over 30 cm) dictates the use of rapid methods, such as analytical models. Traditionally, most industry solutions for ICMV were implemented on hardware circuit boards rather than software to expedite the process. The expensive and lengthy off-project computing time required for numerical solutions prevents them from being practically used in the field to produce real-time ICMVs.

The common mechanism for calculating all types of ICMV involves measuring the vertical acceleration at the center of the vibrating drum and computing ICMV using various models and methods. The concept is simple and ingenious; measuring the properties of compacted materials during compaction allows real-time compaction monitoring and control. The following Fig. B.1 illustrates how ICMV is measured. The roller drum exerts a compaction force on the compacted materials, and the compacted materials react to this force, pushing back against the roller drum. The harder the compacted materials, the larger the reactive force. The accelerometer captures the reaction force in terms of acceleration. The control system will then process the acceleration signals and compute ICMV.

The complexity of the interaction between roller materials during dynamic, vibratory, contact, decoupling, and impact rolling movements makes it challenging to produce accurate measurements. The following is just a brief list of challenges for ICMV:

- Complex Interaction between rollers and compacted materials,
- Difficulty of solving ICMV due to the variances of field measurements, and
- Differences between ICMV and conventional spot tests (footprint size and influence depth)

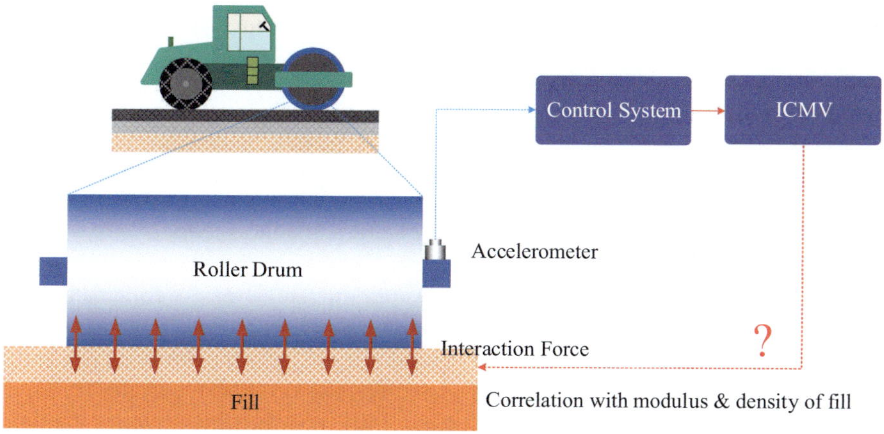

Fig. B.1 Common mechanism of ICMV

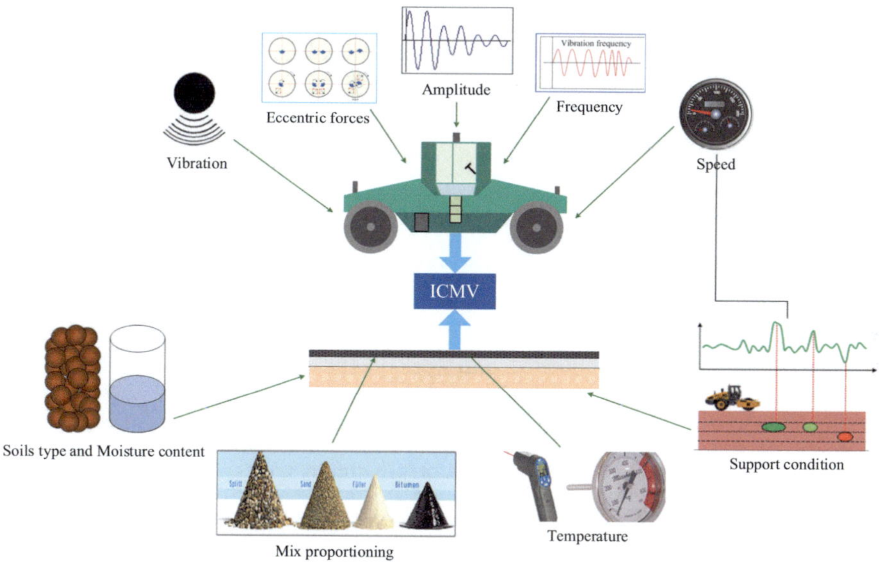

Fig. B.2 The factors that influence ICMV

3 Factors Challenging the IC Measurement Computation

As illustrated in Fig. B.2, numerous factors affect the ICMV computation. On the roller side, the vibration type, eccentric force amplitude and frequency, and roller speed are significant factors. From the compacted materials, the soil type and moisture content, asphalt mixture proportioning, asphalt mix temperature, and underlying support conditions are significant factors.

The following items are the key examples of challenges in producing ICMV:

- *Vibration force is not equal to compaction force*: The vibration force from the eccentric weight in the drum is not equivalent to the effective compaction force that acts on compacted materials. The compaction force fluctuates as the vibration frequencies increase. On the other hand, the vibration force increases monotonically with increasing frequency. This is a common mistake for researchers who begin modeling the interaction between the drum and the material.
- *Actual strain measurement is different from theoretical computation*: The actual strains and displacement under a drum are constant. However, the computed theoretical strains and displacements vary across a drum width based on Lundberg's and Hertz's theories, which are used to solve the contact problem of a cylindrical drum of finite length on compacted materials. The error is due to the incorrect assumption of the theoretical contact area of the models, which mimics the field drum contact condition. Therefore, the theoretical computation needs corrections to match actual field measurements.
- *When decoupled, the computed modulus is erroneous*: When a drum and compacted materials lose contact or are decoupled, the computed modulus is often too low or non-obtainable. The actual modulus should be high, based on the reactive force, compared to the computed values. Therefore, the technique of using the impact model and reactive force can overcome the difficulties of computing ICMV during double-jump movements of the roller drums (Fig. B.3).

4 Models for Solving ICMV

There are various models for solving ICMV, each with varying levels of applicability to address the challenges mentioned above. The list of models is summarized below (see Table B.1). There are five different models with several types of solution methods: dynamic, static, empirical, and mechanistic.

(1) Model A—Empirical Reactive

The reactive models comprise three types of sub-models, as shown in Fig. B.4

- Model A1: Vibratory Frequency Reactive Model is based on the ratio of frequency responses from the reactive force.
- Model A2: Oscillation Frequency Reactive Model is based on areas of "8-shape" plots to reflect the levels of compaction.
- Model A3: Static Rolling Resistance Reactive Model is based on the rolling resistance coefficient of different types of compacted materials and associated machine driver power.

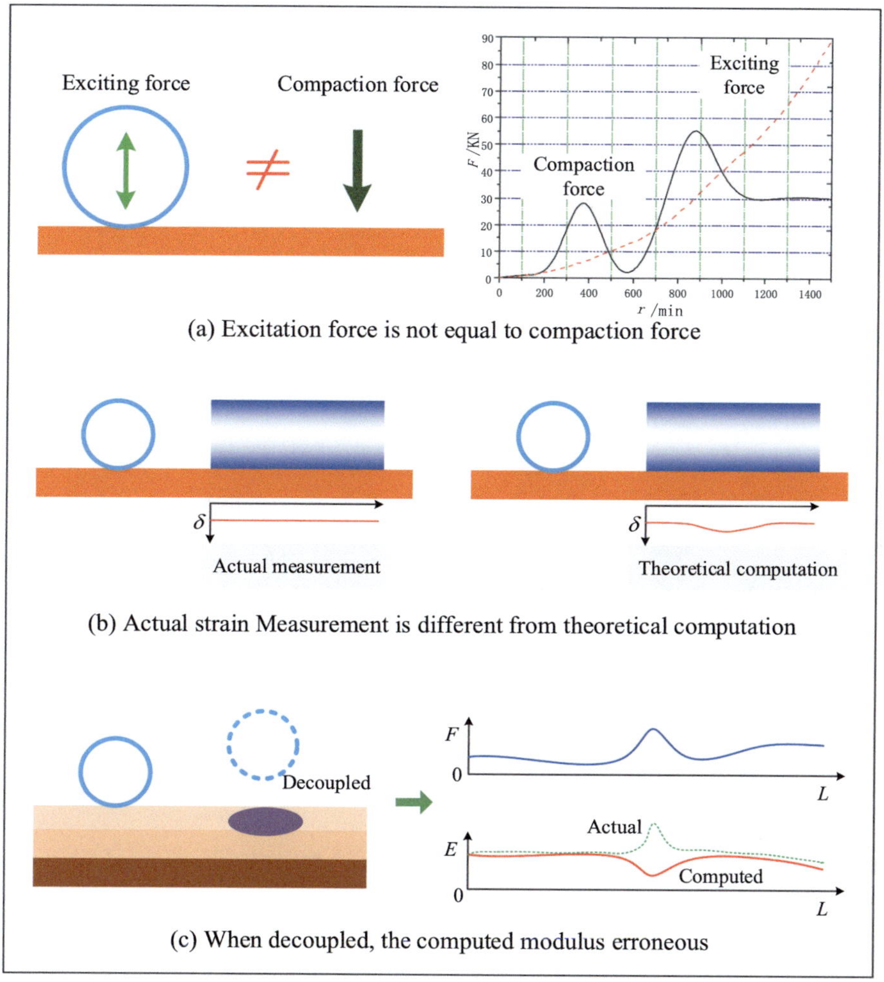

Fig. B.3 Challenges for ICMV solutions

Table B.1 Summary of models for solving ICMV

Model	Description	Mechanistic/ empirical	Dynamic/ static
A	Empirical reactive models	Empirical	NA
B	Continuum roller and half-space layered system	Mechanistic	Dynamic/ static
C	Discrete drum and spring-dashpot coupled system	Mechanistic	Static
D	Dynamic impact model for decoupled drum and layer system	Mechanistic	Dynamic
E	Artificial intelligence method	Mechanistic	Dynamic data

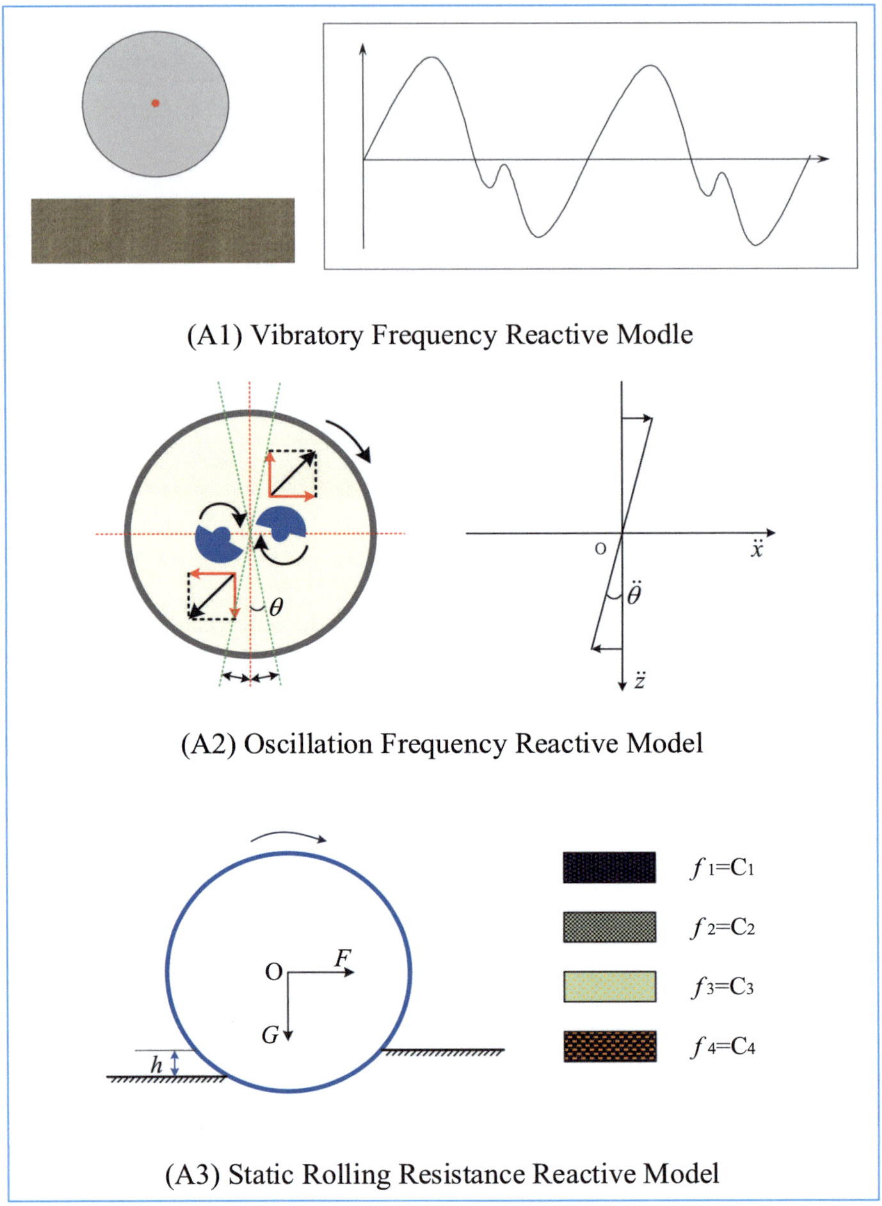

Fig. B.4 Model A—empirical reactive models (with A1, A2, A3)

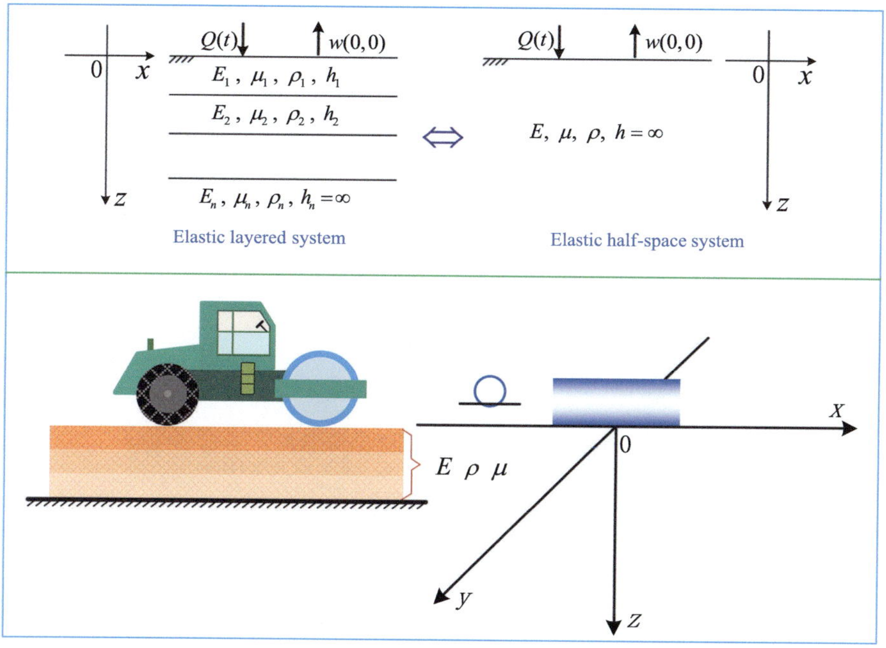

Fig. B.5 Model B—continuum roller and half-space layered system

(2) Model B—Continuum Roller and Half-Space Layered

Using simplification, a pavement layered system can be transformed into a half-space. The compaction force and reactive force are equivalent for both systems. The effective and equivalent modulus, density, and displacement can be computed, see Fig. B.5.

The governing equations for the dynamic method are partial differential equations that are difficult to solve. The static method is simplified by assuming all acceleration terms are zero. However, simplification will depart from the actual condition and require a semi-empirical solution to approximate field measurements.

The flow diagram for computing ICMV with a dynamic model begins from left to right and loops around if the error check is not satisfied (see Fig. B.6). The check is to compute the differences between the actual measured force and the theoretically computed force using the assumed modulus. The analytical solution is challenging because the displacements of compacted materials cannot normally be measured. It requires a tremendous amount of field measurement to provide dynamic correction. Using a numerical solution is also challenging due to the lengthy computing time required. Additionally, back-calculated parameters are not always a unique or exact solution.

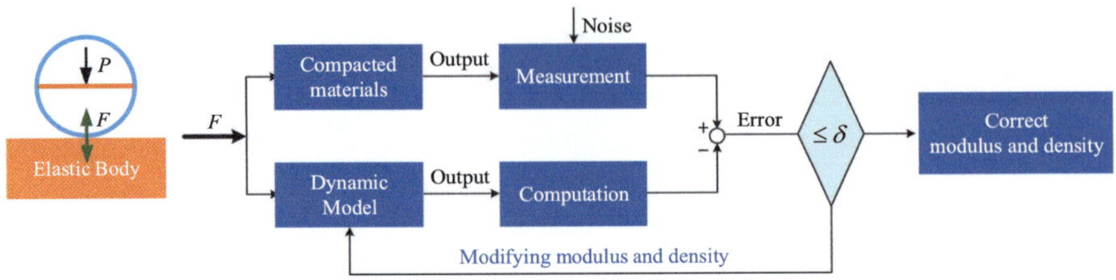

Fig. B.6 Flow diagram for computing ICMV with a dynamic model

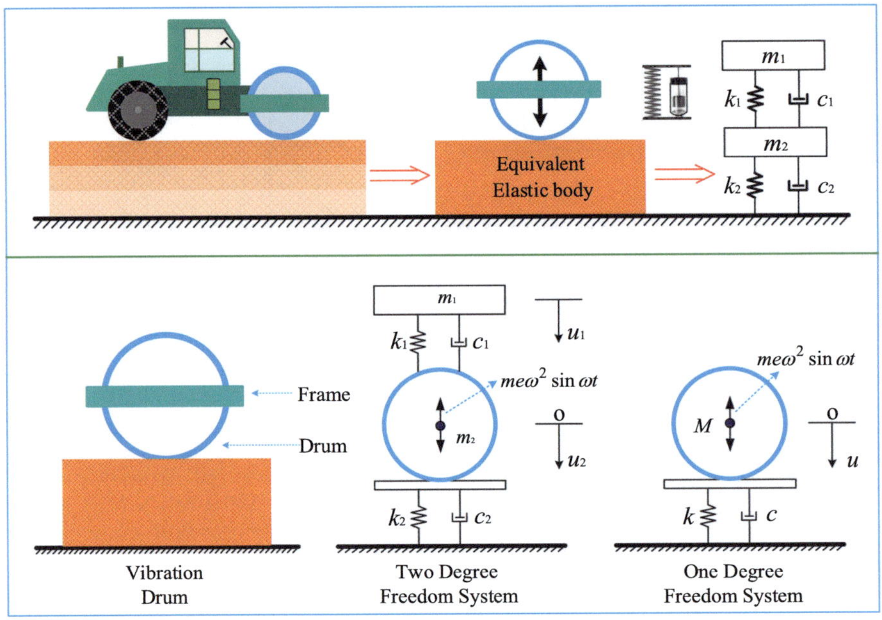

Fig. B.7 Model C—discrete drum and spring-dashpot coupled system

(3) Model C—Discrete drum and spring-dashpot coupled system

Differential equations govern the discrete drum and spring-dashpot coupled system (see Fig. B.7), which is slightly simpler than the solutions for partial differential equations in continuum models. The layer system is simplified to an equivalent elastic body, which is then further modeled as a combination of an elastic spring and a dashpot in parallel, based on the modulus-stiffness conversion theory. A two-degree-of-freedom lump model can be used to model the vibration of the drum (including the frame and drum), or it can be further simplified to a one-degree-of-freedom model. The solution methods include linear and nonlinear ones.

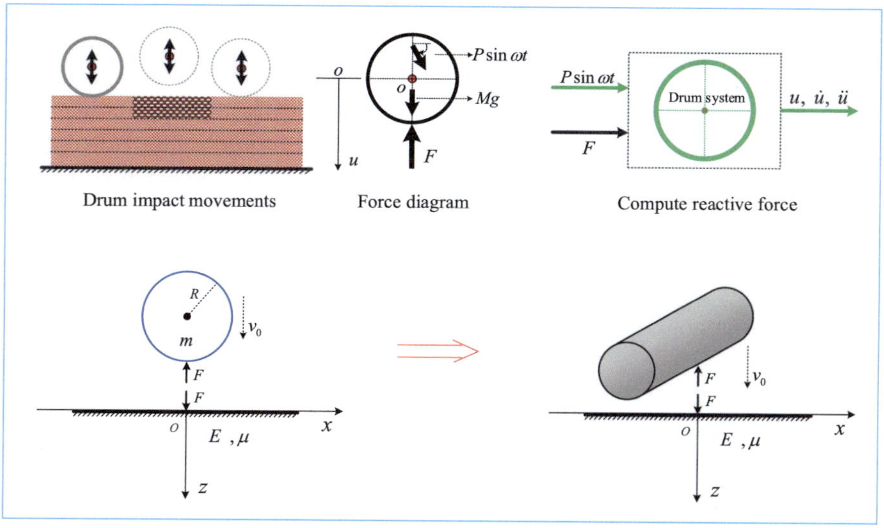

Fig. B.8 Model D—dynamic impact model for decoupled drum and layer system

(4) Model D—Dynamic Impact Model for Decoupled Drum and Layer System

Based on the dynamic impact model for a decoupled drum and compacted layer system, the reactive force and modulus can be computed even when double jumps occur. The drum impact movements can be analyzed with a force diagram to calculate the reactive force. Using a simplified ball impact on a half-space body to emulate the drum impact, the modulus of compacted materials can be calculated using either the displacement method or the impact time method, as shown in Fig. B.8.

(5) Model E—Artificial Intelligence Method

The Model *E* is an artificial Intelligence (AI) model that uses an Artificial Neural Network (ANN) and a Genetic Algorithm (GA). The key is to use sufficient and accurate roller parameters, field measurements of reactive forces, and compacted material properties to train the AI model. The training methods may include Hebbian, Delta rule, and Least Mean Squares (LMS) methods, as shown in Fig. B.9.

The flow diagram for computing ICMV with an artificial intelligence model begins from left to right and loops around if the error check is not satisfied (see Fig. B.10). The check is to compute the differences between the actual measured force and the computed theoretical force with the assumed modulus. The analytical solution is challenging because the displacements of compacted materials cannot normally be measured. It requires a tremendous amount of field measurement to provide dynamic correction.

5 Classification Criteria

The ICMV classification is based on the following five criteria (FHWA, 2017):

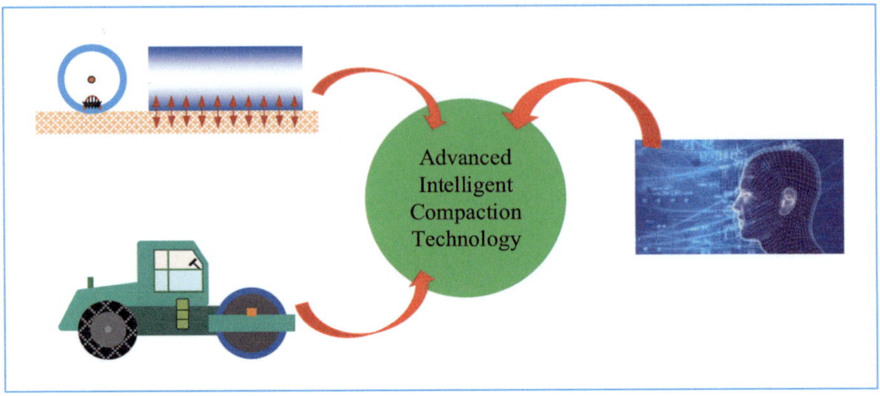

Fig. B.9 Model E—artificial intelligence method

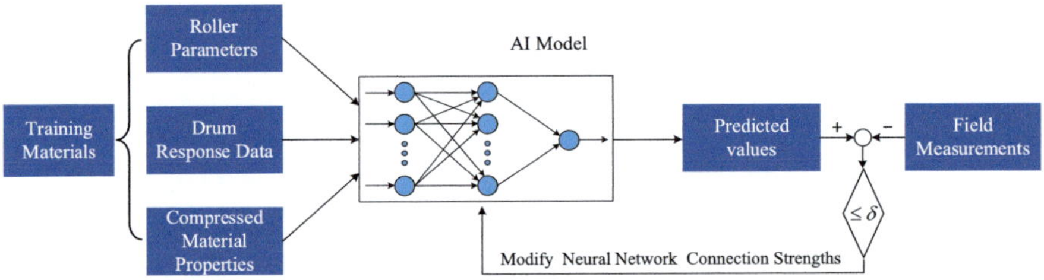

Fig. B.10 Model E—artificial intelligence model

- Correlation with the material's mechanical (modulus) and physical (density) properties.
- Validity during decoupling when a drum loses contact with the compacted materials.
- Capability to allow performance analysis of compacted materials.
- Ability to yield layer-specific mechanical and physical properties of the compacted materials.
- Capability to be enhanced by advanced technology such as AI.

Five levels of ICMV are defined in the subsequent sections, categorized according to these criteria. The ICMV Levels range from 1 to 5, based on whether they meet the above criteria, a mechanistic approach, or an AI-enhanced approach.

6 Classification of ICMV Levels

There are currently solutions for ICMV based on the above models. There is also the next generation of ICMV under development. The following describes the "Levels of ICMV" in terms of applicability to a variety of pavement materials, correlation with material's mechanical (modulus) and physical properties (density), validity during

Table B.2 Five levels of ICMV solutions

Level	Model	Measurement Values	Correlation	Decouple	Layer Specific	Advanced IC
1	Empirical	Harmonic ratio	Satisfactory	No	No	No
2	Energy	Energy index	Unproven	No	No	No
3a	Discrete vibration	Stiffness coefficient	Yes	No	No	Satisfactory
3b	Roller drum movement	Resistance coefficient	Yes	Yes	Satisfactory	Satisfactory
3c	Continuous static	Modulus	Yes	No	Yes	Yes
4	Hybrid	Resistance coefficient modulus	Yes	Yes	Yes	Yes
5	Continuous dynamic	Density, modulus	yes	yes	yes	Yes

Notes

- Correlation: The ICMV can meet correlation requirements with in-situ spot tests. The threshold value for the coefficient of correlation between ICMV and in-situ spot tests is generally accepted as $R \geq 0.70$ or $R^2 \geq 0.5$.
- Decouple: The ICMV can produce a valid solution during a double jump or decouple when the roller drum and compacted material lose contact.
- Layer Specific: The ICMV can produce layer-specific values.
- Advanced IC: The ICMV can be combined with advanced technologies such as artificial intelligence and auto-feedback controls.

decoupling (when a drum loses contact with compacted materials), and applicability to obtain layer-specific mechanical and physical properties of compacted materials. This outlines the roadmap for past, present, and future ICMV development. The evaluation scores of the five levels of ICMVs are summarized in Table B.2.

7 All-Level ICMV Solutions

(1) Level 1 ICMV—Empirical Solutions based on Frequency Responses

The Level 1 ICMVs are empirical solutions based on frequency responses represented by harmonic ratios. These first-generation ICMVs are based on the reactive model for vibratory compaction or the reactive model for oscillation frequency. Examples of Level 1 ICMVs are Compaction Meter Value (CMV), Compaction Control Value (CCV), and HAMM Compaction Value (HMV). The details of CMV and CCV are described below.

① Compaction Meter Value (CMV)

This Level 1 ICMV utilizes the harmonic ratio model, the most widely used ICMV method in the industry. When the vibratory roller acts on soft material or the roller is

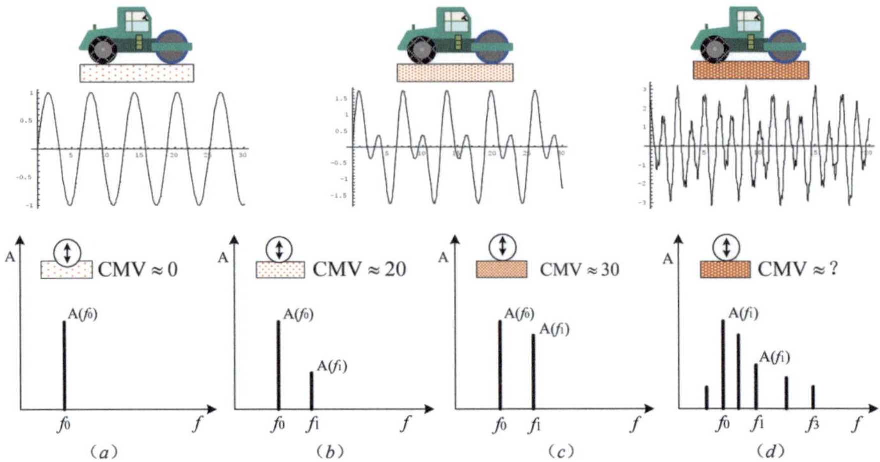

Fig. B.6 The CMV values reflect different compaction levels with frequency responses

suspended in the air, the waveform of the vibration drum response signal is sinusoidal (see Fig. B.6). Only the excitation frequency, f_0, is evident in the frequency-domain response. The waveform will be distorted when the vibratory roller acts on the relatively dense compacted materials. A second harmonic component, f_1, appears in the frequency-domain record in this situation. The ratio of these two frequency signals represents the degree of distortion of the waveform, a.k.a. Compaction Meter Value (CMV). The formula for CMV is shown in Eq. (B.1)

$$\text{CMV} = \text{C}\frac{A(f_1)}{A(f_0)} \tag{B.1}$$

where:

$A\ (f_0)$ is the amplitude at the fundamental frequency;
$A\ (f_1)$ is the amplitude at the first harmonic of the fundamental frequency;
C is a constant.

The harmonic ratio model uses CMV as the ICMV, and its change reflects the change in compaction quality. In Fig. B.6, the compacted material is within a specific stiffness range as the number of rolling passes increases, and the packing state of the compacted material transitions from loose to dense. Under such conditions, the CMV also increases (from Fig. B.6a–c, reflecting the degree of compaction. However, when the compacted materials are very stiff, the changes in the frequency components of the vibration drum response signal appear at multiple frequencies that cannot be ignored (Fig. B.6d). Under such conditions, the CMV cannot correctly reflect the compaction quality information.

② Compaction Control Value (CCV)

Several vendors use harmonic ratio models to address the multiple frequencies observed in Fig. B.6. Examples include various Compaction Control Value (CCV) forms, as shown in Eq. (B.2).

$$\mathrm{CCV} = \frac{A(0.5f_0) + A(1.5f_0) + A(2f_0) + A(2.5f_0) + A(3f_0)}{A(0.5f_0) + A(f_0)} \times 100$$
$$\mathrm{CCV} = \frac{A(0.5f_0) + A(1.5f_0) + A(2.5f_0) + A(3f_0)}{A(2.5f_0) + A(3f_0)} \times 100 \tag{B.2}$$

The harmonic ratio model cannot be applied universally. The field data have shown that the harmonic ratio method can only be applied to limited vibratory rollers compacting materials with specific packing structures. In addition, there is no theoretical proof of why harmonic ratios can be used to monitor changes in the compaction state. Therefore, CMV and CCV should be considered empirical indicators. As in many field studies, the limitations of the harmonic ratio method result in a poor correlation between CMV and conventional spot tests.

It should be noted that, although practices in various countries have shown that CMV does not have universal application value and is often inconsistent with traditional control results. However, there are still many measurement products (CMV + GPS) with CMV as the core, which are mainly used to control the number of times the roller is rolled (including continuous measurement of the rolling temperature of asphalt pavement). Because it is impossible to identify compaction quality parameters, such as modulus, accurately, it is classified as a digital construction category. CMV has given users the impression that intelligent compaction technology is unreliable, which has led to confusion in its application.

No matter how CMV is improved, its approach remains limited to the scope of signal morphology processing and does not address the root of the problem. Practice has proved that this type of method does not have universal application value. In addition, it has not yet been theoretically proven why CMV can be used to reflect changes in compaction status.

(3) **Level 2 ICMV—Energy and Rolling Resistance Solutions**

The Level 2 ICMVs are energy and rolling resistance solutions. It is computed based on the reactive model with rolling resistance for static compaction. Examples of Level 2 ICMVs are OMEGA and MDP.

① Energy Model (QMEGA)

The energy model is more scientific than the harmonic ratio model. It is based on the energy of the interaction between the vibratory roller and the changes in compacted materials during the rolling process. The changes in energy at the end of compaction should be considered constant. The energy model expression that appeared in the 1990s is given in Eq. (B.3).

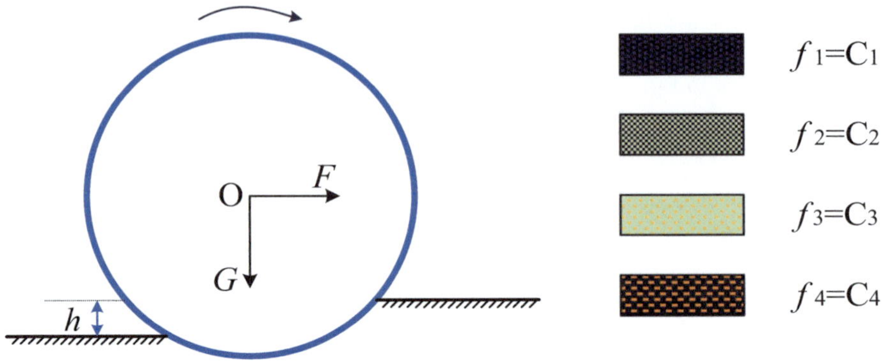

Fig. B.7 Rolling resistance model for static compaction

$$\text{OMEGA} = \oint_{2T} F(u)\dot{u}dt \tag{B.3}$$

where:

F(u) is the resistance force of the compacted materials,
$\dot{u}$ is the speed in the horizontal direction,
t is the time, and T is the vibration period.

This ICMV is referred to as the OMEGA value. This formula is rarely used because the energy method is not directly related to the performance of compacted materials and does not reflect the compaction quality as a direct parameter. Although the expression of the energy model appears scientific, it remains an empirical indicator based solely on the vibration drum response signals.

② Rolling Resistance Model (MDP)

For static compaction, the compaction control can be performed by the phenomenon that "a certain quality static roller can only produce a certain compaction effect." Following this concept, a basic model for tracking the compaction process using a static roller is formed, which can also be considered a continuous compaction control method.

Static compaction can be considered as the rolling of a rigid drum along the surface of a deformed medium. The ratio of the roller's traction force, F (i.e., the relationship between the rolling resistance and the reaction force), and the acceleration of gravity, G, is defined as the rolling resistance coefficient, f. Research and practice have shown that each compacted material has a defined rolling resistance coefficient. During the compaction process, the rolling resistance coefficient changes with the densification of packing, which can be used as an indicator to evaluate the compaction effect (Fig. B.7).

In Fig. B.7, the compression depth (i.e., deformation of the compacted materials) in the rolling process h, produced by the roller compaction of the roller drum, can be divided into two parts: elastic deformation, h_e, and plastic deformation, h_p. The resistance rolling coefficient, f, expression at the start of rolling (considering only plastic deformation, h_p) is as follows.

$$f = \frac{F}{G} = \frac{1}{(1+\mu)(1-\mu/3)}\sqrt{\frac{h}{D}} \tag{B.4}$$

where:

F is roller traction,
G is the acceleration due to roller gravity,
D is roller drum diameter,
h is the depth of depression of the roller drum in compacted materials, where $h = h_e + h_p = h_p$, that is, the elastic deformation is not considered when starting the rolling,
μ is the material condition coefficient, for plastic clay and dry sand, $\mu = 0.5$

The machine drive power (MDP) has been derived according to the above principle. The MDP uses the concept of rolling resistance and compressive deformation to determine the energy required to overcome rolling resistance, as expressed in Eq. (B.5).

$$\text{MDP} = P_g - WV\left(\sin\theta + \frac{a}{g}\right) - (mV + b) \tag{B.5}$$

where:

P_g is the total power required to drive the machine motion;
W is the weight of the roller;
a and g are the acceleration and gravitational acceleration of the machine;
θ is the angle of the roller's forward movement;
V is the speed of the roller;
m and b are the loss factors within the interior of the specific machine, m and b must be calibrated before use.

MDP requires the installation of multiple sensors and the calibration of some parameters, which requires the assistance of the roller manufacturer.

(3) **Level 3 ICMV—Simplified Static Mechanistic Solutions**

Due to its field performance limitations, QMEGA is no longer used. Several studies of MDP have shown that MDP has a shallower influence depth than CMV. Some US DOTs have also observed the MDP's insensitivity to variations in soil stiffness.

The Level 3 ICMVs are based on simplified, static, mechanistic solutions using either discrete vibration, roller drum movement, or static continuum models. As described next, examples of Level 3 ICMVs include stiffness coefficient (discrete vibration model) and modulus (roller drum movement model).

The contents of this section are as follows in Chap. 2 of the original book and will not be repeated. Here's a little clarification on the issue again.

Regarding the stiffness coefficient, there are currently two definitions in use. One is K_s defined based on a discrete model, and the other is K_R defined based on a continuous model.

K_s is based on the lumped linear, mechanistic model to quantify the interaction between the roller drum and the compacted materials. The total force combines elastic and damping forces in the model. The model couples the roller drum with compacted materials, bonded with specific strength, and tightly coupled within a certain vibration intensity range. The model no longer applies when the decoupling happens or the bouncing range is exceeded. Additionally, the mass of compacted materials is not taken into account. Because the volume of the compacted materials changes during roller passes, the density also changes. Therefore, the mass of the compacted materials cannot be determined. Therefore, the discrete model approximates the elastic state toward the end of compaction. However, the vibration balance point changes during plastic deformation causing difficulty in solving the ICMV. The primary concern for compaction quality control is determining the stiffness coefficient of the compacted materials. Although these classic models can be solved, the solution for the stiffness coefficient of the compacted materials in the fields requires sophisticated techniques that are patented or trade secrets. In addition, K_s is greatly affected by roller parameters.

K_R is defined based on a continuous model and represents a resistance coefficient, similar to the Winkler foundation reaction modulus. Its basic characteristics have been introduced in Appendix A. It is a mechanical indicator that is superior to K_s, as it minimizes the influence of roller parameters and has a certain theoretical connection with the modulus. This is an indicator that is different from K_s.

Regarding modulus, although it is based on a continuous model, due to the complexity of the solution, there are currently multiple specific modulus indicators, such as E_{vib} and E_R. In addition, although they can accurately reflect the compaction quality, none of the publicly available formulas can be solved directly. As of now, all specific methods for solving the modulus have not been publicly released.

Level 3 ICMVs provide the best field performance among commercially available ICMVs. However, there are still limitations to Level 3 ICMVs, including the inability to provide reliable measurements during double jumps and layer-specific measurement values.

(4) **Level 4 ICMV—Hybrid Model with Dynamic Mechanistic Solutions**

The Level 4 ICMVs are dynamic, mechanistic solutions that incorporate a hybrid of impact, resistance identification, and modulus estimation models. The commercial versions of Level 4 ICMV are currently unavailable in the US. The following are the backgrounds of Level 4 ICMV.

① Impact Model (Decoupled system of a roller Drum and compacted materials)

The above ICMV models require small elastic deformation, and the surface of the roller drum and the compacted materials must be in close contact and coupled. Otherwise, the models will no longer be valid. Under such a double jump (bouncing) condition, the models cannot provide accurate modulus or stiffness values. Whether the compacted materials are elastic, the roller drum and its surface are not fully coupled. Double jumps often occur during the later stage of compaction in the fields. Thus, it leads to a collision or impact problem. The cause of this bouncing phenomenon is that the rigidity of the compacted materials is inconsistent with the dynamic action of the roller drum. Whether the roller drum bounces are directly related to the modulus, E, of the compacted materials, the vibration mass, M, the excitation force, P, and the frequency, ω. However, it is difficult to determine those parameters theoretically in advance.

If the roller drum bounces, the compacted materials will react to it as an indication of their enhanced resistance to deformation. For a roller with a specific vibration performance, the stiffer the compacted materials are, the stronger the reaction force on the roller drum will be. Therefore, it is more likely that the bounce will occur. The direct performance of the softness and hardness of the compacted materials (i.e., resistance to deformation) is the magnitude of the reaction force, i.e., the resistance force acting on the roller. It has been validated in the field that the resistance force of compacted materials can also be used as a quality control index for compaction. The resistance force can always correctly reflect the change in the compaction state, regardless of whether the roller drum bounces. The modulus and the stiffness coefficient are measures of the ability of compacted materials to resist deformation, which is equivalent to the unit resistance. The resistance force (or the resistance coefficient) applies to uncoupled collision/impact models and the coupled contact models mentioned above.

② Resistance Identification Model

Due to the real-time and continuous requirements of intelligent compaction control, it is impossible to embed sensors in compacted materials to measure their mechanical properties. Therefore, it is necessary to continuously measure the roller drum's dynamic response to identify the resistance force of the compacted materials. Consequently, theoretical solutions and experimental calibration are needed to identify these resistance forces.

To identify the resistance of compacted materials through the roller drum, it is necessary to analyze the dynamic mode of the roller drum's excitation during the compaction process. The resistance force can then be obtained by the system identification method. The movement mode of the roller drum is divided into two types: coupling movement, which occurs in close contact with the rolling surface, and collision movement, which occurs without close contact with the rolling surface, as shown in Fig. B.8a. Regardless of the mode, the effect of the compacted materials on the roller drum can be characterized by the resistance force, F. The force diagram of the steel drum is shown in Fig. B.8b. The structural and performance parameters of the roller drum vibration system are known. Therefore, this collision/impact model has two input parameters: the roller excitation force and the resistance

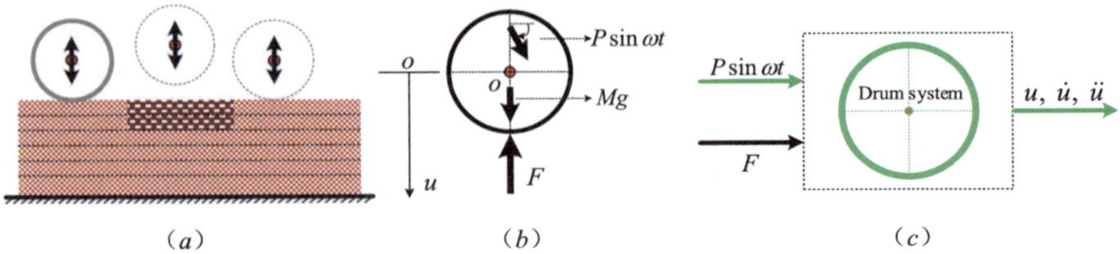

Fig. B.8 Collision/impact model: **a** roller compaction modes, **b** force diagram, and **c** system identification method

of the compacted materials. The output is the dynamic response of the roller drum, as shown in Fig. B.8c.

The rotation of the eccentric block on the roller drum generates the eccentric force. The excitation force is the projection of the eccentric force in the vertical direction and is a known quantity related to the mechanical properties of the roller. The resistance force (F) is an unknown quantity as a function of deformation and rolling. The change of F reflects the change in information about the resisting ability of the compacted materials. The resistance force is also the source of the sinusoidal motion of the roller drum. The roller drum system's output is generally related to the type of sensor used. The commonly used output is the roller drum vibration's acceleration, speed, and displacement. Combined with the test technique and the motion pattern of the roller drum, a resistance identification model based on the system identification method can be formulated as follows.

$$F = Mg + \eta_1 me\omega^2 \sin \omega t - \eta_2 Mf(\ddot{u}, \omega, \varphi) \tag{B.6}$$

where:

η_1 is the dynamic correction coefficient of the excitation force;
η_2 is the comprehensive correction coefficient of the dynamic response of the roller drum;
φ is the phase difference between the excitation force input and the displacement output, which varies with the rolling pass counts and is a function of time series related to the roller drum's dynamic response $\ddot{u}$,, vibration frequency (ω), phase difference (φ), and type of compacted materials need to be determined based on the measured data in the fields.

In practical applications, converting resistance, F, into resistance coefficient F_R (or K_R) will give better results.

③ Modulus Estimation Model

During compaction, the resistance coefficient of the compacted materials, as measured by the roller, is consistent with the variation in modulus. When the roller drum encounters a stiffer area and bounces, the resistance coefficient can still be calculated correctly, resulting in an increased value (see the solid blue line in Fig. B.9).

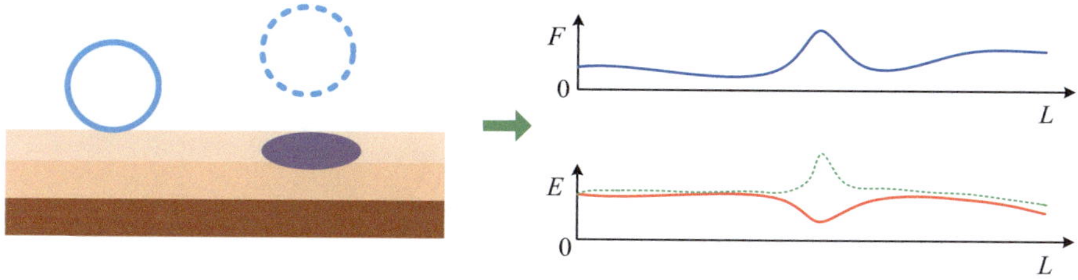

Fig. B.9 Resistance and modulus during double jumps

However, the computed modules are erroneously small. As shown in Fig. B.9, the erroneous moduli (solid red line) should be corrected by the impact model (the dotted green line.

It should be noted that the actual vibratory compaction problem is a mixture of forced vibration and collision impact. The contact time between the roller drum and the compacted surface includes the action time of the forced vibration and the collision impact time. A future study is warranted that would involve nonlinear dynamics. Further details are omitted here due to their complexity.

Since there are no commercially available Level 4 ICMVs, the limitations of their field performance are unknown. Theoretically, Level 4 ICMVs should provide reliable measurements during double jumps and are one step closer to the ultimate Level 5 ICMVs, which can provide layer-specific moduli and densities.

(5) Level 5 ICMV—Dynamic Mechanistic and Artificial Intelligence Solutions

The Level 5 ICMVs combine dynamic mechanistic solutions and artificial intelligence solutions. The commercial versions of Level 5 ICMV are currently unavailable in the US or worldwide. The following are the backgrounds of Level 5 ICMV.

① Dynamic Model

The vibratory roller compaction operation is characterized by vibration and movement during both vibration and rolling. Modeling such phenomena using dynamics is a natural choice. Based on the elastic half-space theory, the dynamic equilibrium, the geometric equation, the compatible equation, and the constitutive equation can be established at any point inside the compacted materials. Further details are omitted here due to their complexity.

② Static Model

The static model is a particular case of the dynamic model. When the acceleration is zero, the dynamics model degenerates into a static one. Currently, the density parameter no longer appears in the model. Only the two parameters of modulus and Poisson's ratio, the critical compaction control index, and modulus still exist.

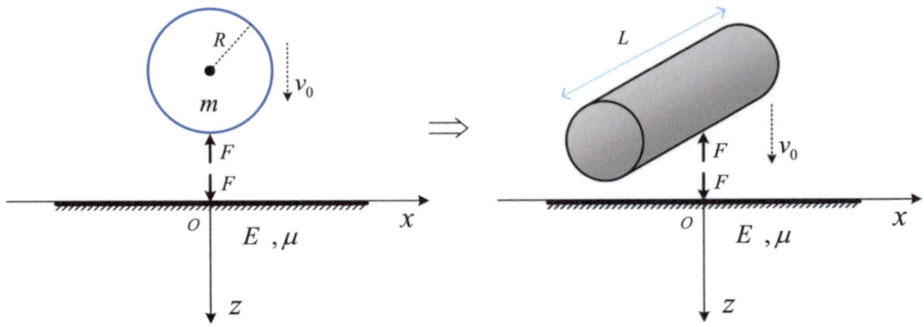

Fig. B.10 Static model for a roller drum and half-space compacted materials

The basic equations of the static model are based on Eqs. (2.1)–(2.4), where the accelerations in the three directions, u, v, and w, are all 0, and the boundary conditions at the surface. However, the compaction force is not the gravitational force of the roller drum; rather, the dynamic compaction force is regarded as static. Therefore, the static model is equivalent to a quasi-dynamic model, which combines dynamic and static forces. The static mode of interaction between the roller drum and the elastic half-space body (compacted materials) is shown in Fig. B.10, where F is the compaction force, L is the drum width, and R is the radius. Further details are omitted here due to their complexity.

③ AI-Enhanced Model

The ultimate ICMV is the Dynamic Model enhanced by Artificial Intelligence (AI). The AI model uses an Artificial Neural Network (ANN) and Genetic Algorithm (GA). The key to success is to use sufficient and accurate roller parameters, field measurements of reactive forces, and compacted material properties to train the AI model. The training methods may include Hebbian, Delta rule, and Least Mean Squared (LMS). The quality of the training data should be emphasized to develop a reliable AI model.

The Level 5 ICMVs can provide layer-specific moduli and density measurements. Layer-Specific Modulus (E_n) and Density (ρ_n) are based on a combination of the Dynamic Impact Model for Decoupled Drum and Layer Systems and the Artificial Intelligence Method. Though still under development, it is envisioned that both models can be fused, and field measurement methods can include layer-by-layer mapping to measure E_n and ρ_n. A true real-time auto-feedback system can then be deployed to optimize compaction without human intervention. Then, an autonomous intelligent compaction machine can be realized, as shown in Fig. B.11.

Since there are no commercially available Level 5 ICMVs, the limitations of their field performance are unknown. Theoretically, Level 5 ICMVs should provide reliable measurements during double jumps and layer-specific moduli and densities.

This appendix explains why IC technology classification is necessary, how to perform IC classification, and provides examples of all IC-level solutions. The primary goal is to dispel several myths about ICs that may hinder industry adoption,

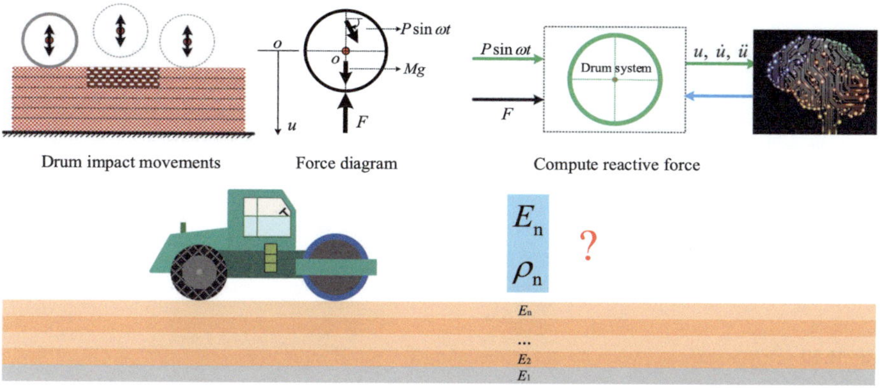

Fig. B.11 Level 5 layer-specific modulus E_{n} and density ρ_{n} and its computation method

thereby limiting their widespread use despite over 50 years of development since the 1970s. However, the IC classification does not discourage low levels of ICs. Each IC level has its application and historical significance. The IC is still evolving through its 50-plus years of development and adoption. It is anticipated that more innovations will extend the IC measurement mechanism and encourage widespread application in the future.

8. **Further Discussion of Related Issues**

We have discussed the classification of intelligent compaction technology in detail above. However, this classification was completed in 2017. Since the understanding of intelligent compaction was not sufficiently deep at that time, the classification given above primarily reflected the perception technology in intelligent compaction, and it was based on the control index. Strictly speaking, intelligent compaction should also encompass "analysis, decision-making, and execution," as well as other components; that is, the complete classification should include the classification of "perception, analysis, decision-making, and execution." Currently, the "analysis, decision-making, and execution" aspect has been integrated into the technical classification of construction automation. Therefore, the classification of intelligent compaction technology remains applicable to its development until 2017. However, some issues require further explanation and discussion.

(1) Regarding the control index of intelligent compaction technology above Level 3

The core of early continuous compaction control lies in selecting the control index problem, which is crucial for evaluating compaction quality. According to the technical classification of 2017, Level 3 technology provided three mechanical control indicators: modulus E, resistance coefficient K_{R}, and stiffness coefficient K_{S}. The mechanical models corresponding to these three indicators are different, as introduced above. Here are some additional explanations.

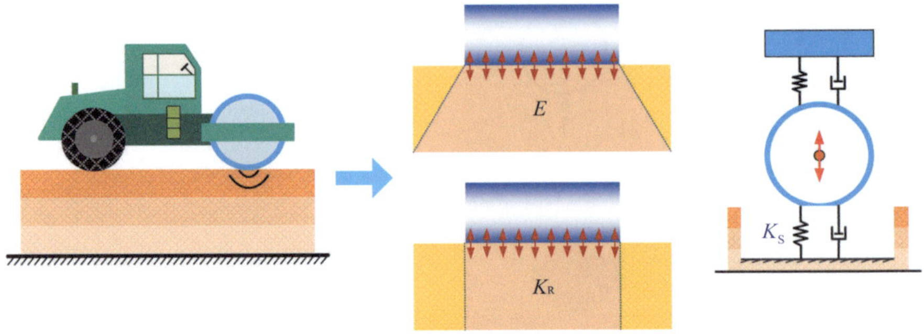

Fig. B.12 Differences between the three mechanical control indicators

Modulus E appears in the continuous model and is the most important physical parameter in the continuous medium. When the compacted body is regarded as a continuous medium, the modulus is, of course, the best choice as an indicator for evaluating its mechanical properties. This is also the primary reason why the modulus is selected as the indicator for intelligent compaction control. However, the difficulty of solving the modulus is also the greatest. So far, all solution methods have not been made public, and all solutions are approximate, although they appear to be analytical solutions.

The resistance coefficient, K_R, is also referred to as the stiffness coefficient, based on the roller parameters. Although it also appears in the continuous model, it is an indicator established in imitation of the Winkler foundation reflection modulus. It only considers the vertical deformation within the range of the steel drum (equivalent to the existence of many vertical springs). This is different from the modulus, but there is a certain theoretical connection between K_R and E, which can be determined through theoretical derivation.

The stiffness coefficient Ks is also referred to as the spring coefficient, which corresponds to the discrete model (where the compacted body is regarded as a discrete body composed of springs and damping), thereby ignoring the mass of the compacted body involved in the vibration. When the spring coefficient Ks is used as a control index for compaction quality, the damping of the compacted body is ignored.

From the perspective of quality control, although the definitions of E, K_R, and Ks differ, these three indicators can accurately reflect changes in the mechanical properties of the compacted body and can be used as indicators for compaction control. This is also the primary reason why these three indicators are featured in Level 3 technology. However, it is also worth noting that there are significant differences between these three indicators. Figure B.12 shows the differences between them.

In Fig. B.12, both control indexes E and K_R regard the compacted body as a continuum, which is in line with the actual field condition. The modulus E corresponds to the elastic half-space model in elastic mechanics (or soil mechanics). The compaction force of the roller is diffused downward, while the stiffness coefficient K_R only considers the propagation of the compaction force within the vertical range of the steel drum. This idea comes from the Winkler foundation reaction modulus. Suppose the compacted body is an ideal elastic body. In that case, there is a theoretical connection between the two indices (see the definition provided in this book), and their dimensions are both MPa (MN/m^2), but their meanings remain different. Modulus E means the size of the load that the continuum can bear per unit area, and the stiffness coefficient K_R means the size of the linear load when the continuum produces unit deformation under the action of the steel drum (this definition basically eliminates the influence of the steel drum size). When performing mechanical analysis on the compacted body (which is often encountered during the maintenance period), the modulus is undoubtedly the best mechanical performance index. If it is only for compaction quality control, both indicators are feasible.

The early stiffness coefficient K_S (spring coefficient) was obtained by treating the compacted body as a discrete body composed of springs and damping blocks. It refers to the size of the load under unit deformation, and its unit is KN/m. The discrete body model is primarily used in the field of vehicle dynamics to simulate the reaction of the compacted body (roadbed and pavement, railway subgrade) to the vehicle. However, it cannot be applied to the mechanical analysis of the compacted body itself. In addition, K_S is significantly influenced by the size of the steel drum, and the Ks obtained with different rollers also vary. Despite this, it is still feasible to use K_S for compaction quality control.

In the classification of intelligent compaction technology, Level 3 technology marks the watershed between mechanical indicators and empirical indicators. It also serves as the starting point of advanced intelligent compaction technology. Although the definitions of the three mechanical indicators differ, they can all be used for compaction quality control and accurately reflect compaction quality information, which is distinct from the empirical indicators.

It should be noted that the correlation between these mechanical indicators and the results of traditional point mechanical tests is sometimes not very good. Still, the mechanical indicators themselves do not cause it, but rather the unevenness of the filler and the different test sizes (see the analysis in Appendix A for details), which differ from the empirical indicators.

Regarding Level 4 technology, the primary difference between it and Level 3 technology is that it can address the bounce issue of the roller. The essence of the bounce problem (i.e., double-jump) is that the steel drum of the roller is not in continuous contact with the rolling surface, which will cause the displacement of the steel drum to be inconsistent with the deformation of the rolling surface (see Fig. 4.6). The displacement of the rolling surface cannot be estimated based on the displacement of the steel wheel, which renders the continuum model inapplicable. Based on experience, there are two ways to address this problem. One approach is to utilize the impact model, as explained above; the other is to employ the continuous model and resolve the bounce issue of the steel drum through data processing technology.

Regarding Level 5 technology, its core is to achieve density and modulus simultaneously through the dynamic model. This is a very difficult problem. See the discussion below for details.

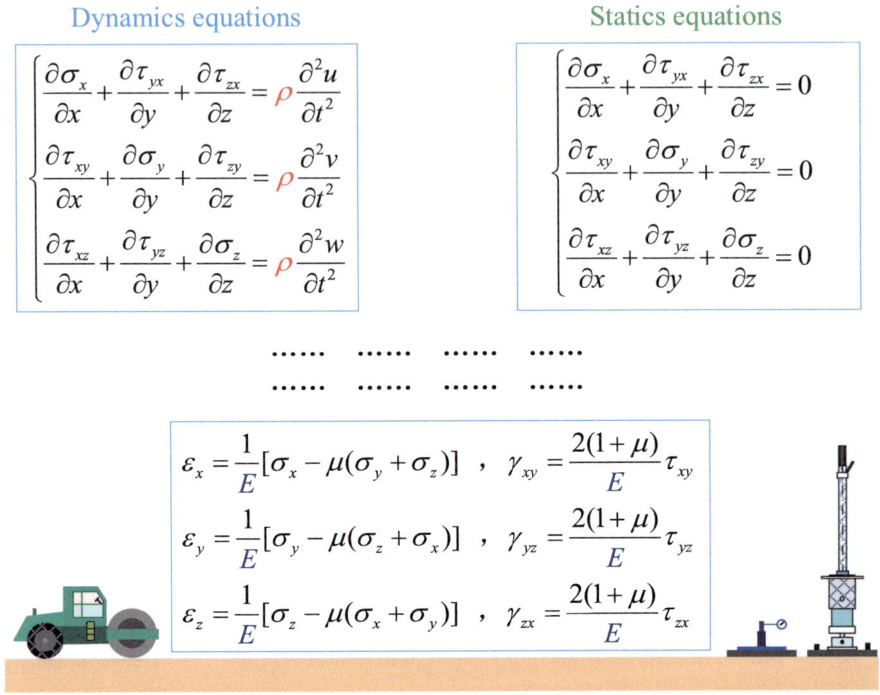

Fig. B.13 Dynamic model and static model of landfill

(2) Further discussion on density and modulus in mechanical models

Density and modulus are two independent indicators that characterize the physical and mechanical properties of continua (i.e., fill). The uses and methods of obtaining these two indicators have been discussed in the relevant chapters of this book and will not be repeated here. Here, we mainly discuss the problem of obtaining density and modulus in the mechanical model of continua, which is the core of Level 5 technology.

When the filler is regarded as a continuum, its mechanical model is represented by a partial differential equation, which includes equilibrium equations, geometric equations, strain coordination equations, and constitutive equations. For details, see Chap. 2 of this book. The mechanical model of continua is primarily divided into dynamic and static models. The primary difference between the two is whether the influence of acceleration is taken into account. The dynamic model considers the influence of acceleration in the equilibrium equation, and the density parameter appears. In the static model, there is no density in the equilibrium equation. However, the geometric equation, strain coordination equation, and constitutive equation remain unchanged from those in the dynamic model. See Fig. B.13.

For the dynamic model of the continuum, since the two parameters, density and modulus, appear simultaneously in the corresponding partial differential equation

(Posson's ratio μ is also a parameter, but because the variation is not large, it is generally assumed to be a constant), this provides the conditions for simultaneously identifying density and modulus. If an analytical solution can be obtained (such as the displacement solution of the continuum surface), its expression must include the two parameters of density and modulus. In theory, the density and modulus can be inversely calculated based on the solution expression to satisfy the dynamic equation. Although this idea is correct in theory, the actual field condition is complicated. Let's discuss it briefly below.

Calculating the parameters of the equation based on the solution is an inverse problem in mathematics. If there is only one parameter, the solution obtained is generally unique. For example, in the static model, only the modulus parameter is unknown. The expression of the modulus can be obtained based on the surface displacement solution (analytical expression) (see Sect. 2.4 of this book). In this way, the modulus can be determined based on the plate load test (Fig. B.13) data (measured data of force and displacement). This is the traditional method of determining the modulus. In the dynamic model, there are two unknown parameters, density and modulus. However, since there is no complete analytical solution, it is impossible to obtain the expressions of density and modulus based on the solution. In this case, a parameter identification method is necessary to address the issue, which leads to the problem of multiple solutions.

To determine the modulus and density using the parameter identification method, a set of estimated values of density and modulus must be given in advance, and then iterative calculations must be performed until the requirements are met. For specific operations, please refer to references [1, 3]. The main problem is that multiple solutions (combinations of multiple densities and moduli) can meet the requirements, and it is necessary to make choices based on experience, which is somewhat difficult. In addition, since the dynamic equation has not yet been solved analytically, it can only be solved using numerical calculation technology, which requires a large amount of computation and a long calculation time, making it difficult to meet the needs of real-time control on-site.

In summary, although the dynamic model of the continuum measures density and modulus simultaneously, current technology still makes it difficult to obtain these two parameters in real-time during the rolling process. There is still a long way to go to achieve Level 5 intelligent compaction technology.

(3) On the relationship between intelligent compaction technology classification and automated construction classification

The intelligent compaction technology classification is primarily used for compaction quality control, while the construction automation classification encompasses the entire construction process (here, it mainly refers to filling engineering). The two are both different and related. Let's discuss them briefly below.

The intelligent compaction technology classification was proposed in 2017. Due to the limitations of the technical conditions at the time, the four basic characteristics of intelligent compaction (perception, analysis, decision-making, and execution) had

Fig. B.14 Intelligent compaction technology classification and automated construction classification

not yet been proposed. It was proposed solely from the perspective of perception technology, with its core being the classification of control indicators. Now it appears that this classification has some incompleteness. Still, the automated construction classification addresses this shortcoming. It considers the four basic characteristics of intelligent construction, namely "perception, analysis, decision-making, and execution" (see the discussion in Appendix A1 for details). Strictly speaking, the intelligent compaction technology classification should be included in the classification of automated construction. Figure B.13 shows the corresponding relationship between the two. Among them, the Level 3 technology of intelligent compaction is advanced and is the key to realizing the Level 2 technology of automated construction. The following is some explanation of Fig. B.14.

In the classification of intelligent compaction technology, advanced intelligent compaction technology starts from Level 3. The control indicators of Level 1 and Level 2 are empirical indicators and can be classified as the same level. This is why these two levels of technology correspond to L1 technology in automated construction. Among them, Level 2 intelligent compaction technology has not been widely adopted and has been largely discontinued. Instead, Level 1 technology, represented by harmonic ratio indicators, has occupied a large portion of the application market. The primary reason is that the harmonic ratio principle is relatively simple and lacks patent protection. However, few people use it for compaction quality control, but only as an auxiliary indicator for counting the number of rolling passes. The advanced intelligent compaction technology is represented by the mechanical control indicator (Level 3), which corresponds to the L2 technology in automated construction.

According to actual investigations, the current automated construction, L1 technology is primarily composed of the automatic driving technology of engineering machinery (mainly rollers) and the compaction technology represented by CMV, which enables automatic rolling. Its control target is the specified number of rolling passes and the planned compaction route. L2 technology and above are all based on Level 3 intelligent compaction technology, which requires the correct perception of compaction quality information during the automatic rolling process. On this basis, the "analysis, decision-making, and execution" parts are also stipulated. For details, see the relevant discussion in Appendix A.

(4) About the Information Platform

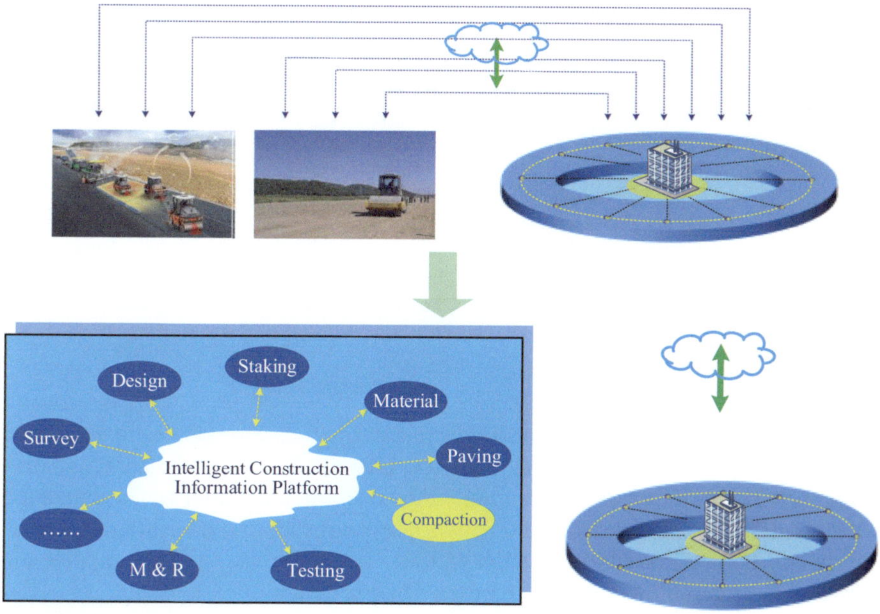

Fig. B.15 Relationship between the intelligent compaction information platform and the intelligent construction information platform

Intelligent compaction is carried out during the rolling process. The compaction control system installed on the roller can display the corresponding information perceived in real time. Although the content displayed by different levels of intelligent compaction technology varies, it is generally necessary to display, analyze, and manage various information perceived on-site on the background computer. The computer and its corresponding software that undertake this task are the compaction information platform, which is a component of the Intelligent Construction Information Platform, as shown in Fig. B.15.

In the past, after rolling was completed, the compaction control system would transmit various information (data) perceived on-site to the background computer through wired communication. The compaction information management software in the computer would then perform secondary display, analysis, and archiving to generate a compaction quality report. With the advancement of science and technology, particularly the widespread adoption of network technology, it has become possible to transmit information through wireless networks, allowing for real-time transmission. When the roller is equipped with a compaction control system that has this function, the engineering Internet of Things is realized, and the corresponding background computer and compaction information management software are also upgraded to the level of the compaction information platform, as shown in Fig. B.15.

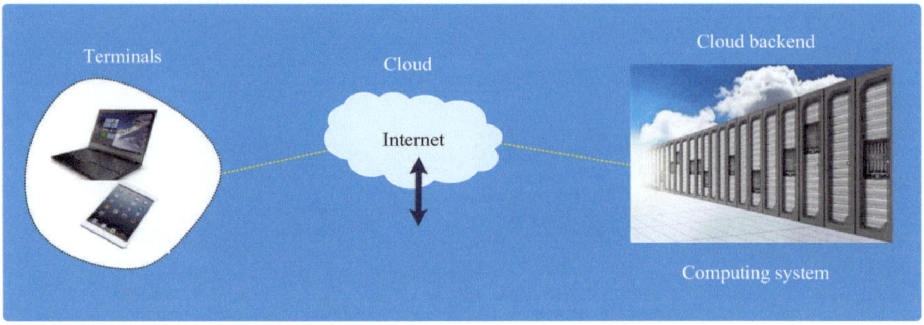

Fig. B.16 Meaning of cloud platform

The information platform is a fashionable term, and its essence lies in the digital and networked existence of information. In the information age, understanding information is typically based on the meaning of the information platform. Digitalization corresponds to computerization, and networking corresponds to the Internet. Therefore, in layman's terms, an information platform is a combination of a computer and a network, a system that combines hardware and software.

There are many types of information platforms, among which the cloud platform (Fig. B.16) is the most mainstream. "Cloud" (also a widely used term) is a metaphor for the Internet, and its concept originates from "cloud computing". A cloud platform can realize the interconnection between various terminal devices. All resources enjoyed by users are provided by a cloud backend (a computer system, i.e., a cloud platform) with super storage and computing power. An intelligent construction information platform containing compacted information is a cloud-based platform.

In intelligent construction, the information platform serves as a bridge between subsystems and users. All terminals (including subsystems and users) exchange information with the platform—passing various data to the information platform, such as design information, construction information (materials, progress, quality, etc.) and various test data, etc., which are uniformly processed by the platform and then various instructions are issued to the terminal (see Fig. B.15). This is also one of the foundations for realizing digital twins and virtual construction (see the introduction volume of the series for details).

The current intelligent construction information platform is still in its preliminary stages and lacks significant intelligence. Therefore, it may be more accurately described as an engineering construction information platform. This type of platform contains very rich content (Fig. B.15). The compaction information platform is one of the subsystems. It not only solves the problem of wireless transmission of on-site compaction data, but also conducts a comprehensive analysis of compaction data and automatically generates quality reports. Some platforms can also facilitate information interaction between the platform and the site, enabling remote monitoring of compaction quality. Currently, the countries with engineering information platforms are primarily China and the United States.

In China, research and development of engineering information platforms began in 2013, under the auspices of the Engineering Management Center of the Ministry of Railways at the time. The information platform includes bridges, tunnels, lines, stations, and other parts, and compaction is only one of the modules (Fig. B.15), which requires real-time transmission of on-site rolling data. Currently, the platform has been fully promoted and implemented in the railway construction system nationwide. In addition, with the promotion of the railway engineering construction information platform, China's highway field has also developed many similar engineering information platforms since about 2017. These platforms, mainly organized and developed by the highway construction authorities of various provinces, have been applied in their respective highway construction projects.

In the United States, compaction data management is primarily handled by the Veta platform, as specified in several national standards of the American Association of State Highway and Transportation Officials (AASHTO). Veta is a geospatial software tool for Intelligent Construction Data Management (ICDM). Veta was first developed in 2012, funded by the US Federal Highway Administration (FHWA) and the Minnesota Department of Transportation (MnDOT), to import, display, filter, and analyze IC data. Since then, Veta has evolved to include paver-mounted thermal profiling (PMTP) and dielectric profiling system (DPS) data. Therefore, Veta is a public domain software that can be freely downloaded, installed, and used by anyone. The Transportation Pooled Fund (TPF)-5(466) National Road Research Alliance (NRRA) supports and funds the ongoing enhancements to Veta, keeping it free to the public.

The Veta desktop version has run on MS Windows or an equivalent operating system since 2013. Veta is migrated to the cloud platform, Veta Center (www.VetaCenter.com), that includes Veta Material Delivery Management System (MDMS) and Web, in 2025 (https://www.intelligentconstruction.com/projects/veta-mdms-and-web/). Veta's functionality includes viewing and analyzing intelligent construction technologies (ICT), such as IC, PMTP, and DPS data. Alignment files are optional for importing into Veta, allowing for easy viewing and filtering. Spot test data can also be input to Veta for correlation analysis. ICT data can be displayed on various site maps for visual inspection. Data filtering is a powerful feature of Veta that extracts the desired data for viewing and analysis. Veta can perform various statistical analyses for the overall data and sublots and then create reports in various forms. Veta is important in facilitating the ICT standardization and the agency's implementation. Standardization is crucial for the adoption and implementation of any innovative technology. Veta is also useful for contractors' quality control (QC) and meeting quality assurance (QA) requirements. Although Veta is developed in the US, its users are scattered worldwide, with the majority residing in the US, Europe, and Australia. Veta desktop software can be freely downloaded at www.intelligentc onstruction.com.

References

1. Xu, G. (2016). *Principles and engineering applications of continuous compaction control of roadbed*. Science Press.
2. [US] Lucci, S. (2019). *Artificial intelligence* (C. Lin, Trans.). People's Posts and Telecommunications Press.
3. Xu, G. (2019). *Continuous and intelligent compaction control technology for high-speed railway subgrade* (p. 2019). China Railway Publishing House.
4. [Australia] Negnevitsky, M. (2015). *Artificial intelligence—guide to intelligent systems* (W. Chen, Trans.). Mechanical Industry Press.
5. [Canada] Sutton, R. S. (2019). *Reinforced learning* (K. Yu, Trans.). Electronics Industry Press.
6. Wei, W. (2018). *Intelligent control technology* (p. 2018). Mechanical Industry Press.
7. [US] FHWA. (2017). *Intelligent compaction measurement value (ICMV)—A roadmap (FHWA-HIF-17-046)*. FHWA.

G. Xu et al., *Pioneer of Intelligent Construction - Intelligent Compaction*,
https://doi.org/10.1007/978-981-95-4184-3

MIX
Papier aus verantwortungsvollen Quellen
Paper from responsible sources
FSC® C105338

If you have any concerns about our products,
you can contact us on
ProductSafety@springernature.com

In case Publisher is established outside the EU,
the EU authorized representative is:
Springer Nature Customer Service Center GmbH
Europaplatz 3, 69115 Heidelberg, Germany

Printed by Libri Plureos GmbH
in Hamburg, Germany